Local Emergency Planning Committee Guidebook

UNDERSTANDING THE EPA RISK MANAGEMENT PROGRAM RULE

This is one in a series of CCPS *Concept* books.
A complete list of publications available from CCPS appears at the end of this book.

Local Emergency Planning Committee Guidebook

UNDERSTANDING THE EPA RISK MANAGEMENT PROGRAM RULE

A CCPS CONCEPT BOOK

R. J. Walter
AntiEntropics Inc.
New Market, Maryland 21774

CENTER FOR CHEMICAL PROCESS SAFETY
of the
American Institute of Chemical Engineers
3 Park Avenue, New York, New York 10016-5901 USA

Library of Congress Cataloging-in-Publication Data
Walter, R. J., 1957–
Local emergency planning committee guidebook : understanding the EPA Risk Management Program Rule / R.J. Walter.
p. cm.
Includes bibliographical references and index.
ISBN 978-0-8169-0749-6
1. Hazardous substances—Safety measures—Handbooks, manuals, etc.
2. Hazardous substances—Risk assessment—Handbooks, manuals, etc.
I. American Institute of Chemical Engineers. Center for Chemical Process Safety II. Title.
T55.3.H3W35 1998 98-44789
363.17′3—dc21 CIP

Contents

Preface

The Center for Chemical Process Safety (CCPS) was established in 1985 by the American Institute of Chemical Engineers (AIChE). It exists for the express purpose of assisting chemical and hydrocarbon industries in avoiding or mitigating catastrophic chemical incidents. This effort was expanded in recent years to include all facilities that handle chemicals.

Local emergency planning committees (LEPCs) play a key part in effective implementation of risk management programs (RMPs). The Clean Air Act 112 (r) rule addressed by this concept book shows how EPA has increased the scope for LEPCs from just responding to releases to taking a proactive role in helping facilities in their communities prevent releases.

LEPCs can be a central point around which emergency management agencies, responders, industry and the community can gather to work together to find solutions to risk management issues. LEPCs may play an active role in RMP-related activities such as

- risk communication,
- public education,
- industry outreach,
- mitigation, and
- emergency planning.

The Emergency Planning and Community Right-to-Know Act (EPCRA) that established LEPCs tasks them with enhancing hazardous materials safety through education, emergency planning, response training, coordinating exercises, and critiquing responses to actual releases. LEPCs are now expected to research and question the meaning of information they receive under the RMP program.

This book is intended to provide information to LEPCs to help these groups address the issues that affect process safety in their areas.

Some LEPCs have gone further than RMP rule requirements explicitly require to assist their communities in being better prepared. As an example, LEPCs from the Kanawha Valley area of West Virginia and Augusta, Georgia have already helped present and explain RMPs to the public. Other LEPCs are helping to inform potentially regulated facilities that they may be required to submit an RMP to the EPA on or before June 21, 1999.

Not all LEPCs have the same resources or concerns. Some LEPCs exist in name only while others maintain constant dialogue with the chemical processing facilities in their response areas to stay up-to-date on changes in potential response needs. The RMP rule provides a driving force to help awaken some "sleeping" LEPCs.

Regardless, LEPCs have unlimited opportunity to be involved in their community's risk management and even take charge of some aspects. This concept book presents several suggestions for using the information that will become available to your LEPC and some methods to help industry and your community maintain a high standard for their risk management programs. Though this text is intended primarily for LEPCs, it would also be very useful to chemical facility personnel who may not be familiar with this regulation.

Acknowledgments

The American Institute of Chemical Engineers and the Center for Chemical Process Safety thanks the members of the Small and Medium Enterprises Subcommittee for their dedicated efforts and technical contributions to the preparation of this text. CCPS also expresses appreciation to the members of the Technical Steering Committee for their advice and support

The chair of the Small and Medium Enterprise Subcommittee was Eric J. Lohry of Nutra-Flo Company. The Subcommittee members were John Hudson, formerly with PCR Company; Homer E. Kunselman, Dow Corning Corporation; Linda Hicks, Reilly Industries, Inc.; Claud Eley, Clark Oil & Refining Corporation; and Douglas Heel, Westinghouse Corporation. Ray E. Witter was the CCPS Staff Liaison and was responsible for the overall administration and coordination of the project.

The members of the Small and Medium Enterprises Subcommittee would like to thank their respective employers for providing time to participate in this project

Robert Walter of AntiEntropics Inc. was the author of this text and editing was provided by AntiEntropic's Sandra Baker.

AntiEntropics and CCPS would also like to acknowledge the following groups and individuals for their assistance in researching and developing the information presented in this book: Mr. Rick Foottit, Conoco Inc.; Mr. Joe Dooley, Exxon U.S.A.; Mr. Greg Brown and Mr. Pat Kimmet, Cenex Refining; and Chief Lorren Ballard of the Billings Montana Fire Department and Chairman of the Yellowstone County LEPC.

CCPS also gratefully acknowledges the comments and suggestions submitted by the following peer reviewers: Mark

French, PPG Industries, Inc.; Robert Hammann, Amoco Chemical Corporation; John Hudson, Joe Kulak, Novartis Corporation; and Arthur G. Mundt, The Dow Chemical Company. Their insight and thoughtful comments helped ensure an accurate and up-to-date text.

Acronyms

AIChE American Institute of Chemical Engineers
AIHA American Industrial Hygiene Association
ANSI American National Standards Institute
API American Petroleum Institute
ARS Alternative release scenario
ASTM American Society of Testing and Materials
BLEVE Boiling liquid expanding vapor explosion
CAA Clean Air Act
CAER Community awareness and emergency response
CAS Chemical Abstract Service
CEPPO Chemical Emergency Preparedness and Prevention Office
CMA Chemical Manufacturers Association
CFR Code of Federal Regulations
DOT Department of Transportation
E&P Exploration and production
EHS Extremely hazardous substance
EPA Environmental Protection Agency
EPCRA Emergency Planning and Community Right-to-Know Act
ERP Emergency response program
ERPG Emergency Response Planning Guideline
FMEA Failure mode and effects analysis
FR *Federal Register*
LFL Lower flammability limit
LOC Level of concern

LEPC	Local emergency planning committee
NAICS	North American Industrial Classification System
NFPA	National Fire Protection Association
NIOSH	National Institute for Occupational Safety and Health
NRT	National Response Team
NWS	National Weather Service
OCA	Offsite consequence analysis
OCS	Outer continental shelf
OSHA	Occupational Safety and Health Administration
P&ID	Piping and instrument diagram
PSM	Process safety management
RCRA	Resource Conservation and Recovery Act
RMP	Risk management program
RMP rule	Risk management program rule
RMPlan	Risk management plan
RTK	Right-to-Know
SARA	Superfund Amendment and Reauthorization Act
STEP	Strategies for Today's Environmental Partnership
TNT	Trinitrotoluene
USGS	United States Geological Survey
VCE	Vapor cloud explosion
WCS	Worst-case scenario

1

Introduction

Local Emergency Planning Committees as Risk Management Program Stakeholders

As day-to-day reliance upon benefits from advanced technology has increased, some hazards due to our increased capabilities have presented themselves. Technology has proven that it can bite back in various ways. In the computer software and hardware industry, viruses can destroy our ability to use some of our most powerful modern tools. In the telecommunications industry, equipment malfunctions can interfere with our increasingly complex communications systems. However, in the chemical processing industry sometimes the hazards have been realized with catastrophic results. Some historic toxic and flammable releases that illustrate this are listed below:

Year	Location	Material
1974	Flixborough, UK	Cyclohexane
1976	Seveso, Italy	Dioxin
1984	Mexico City, Mexico	Liquefied Petroleum Gas
1984	Bhopal, India	Methyl isocyanate
1988	Norco, LA	Propane
1989	Pasadena, TX	Ethylene and Isobutane

Each of the above releases caused great losses. When something goes wrong with our technology during the processing of hazardous chemicals, either in design, construction, or operation, dramatic effects can result in multiple fatalities or injuries.

These types of releases stirred the government to respond with regulatory prevention measures. Since 1985 some important regulations have been established in the United States to protect citizens from the effects of toxic, flammable, and explosive substances. Government recognized that there are many types of stakeholders involved in this effort:

- regulators,
- state and local government,
- industry,
- environmental groups,
- response organizations, and
- the public.

The likelihood that we will end our addiction to the benefits of technology any time soon is slim. However, more intelligent use of our technological resources and our ability to organize and control them can reduce the risks we face.

The following list of regulations and official programs shows governments' recent trend toward enforcing intelligent use and control of our capabilities to prevent and respond to chemical risks:

Year	**Regulation**
1985	Chemical Emergency Preparedness Program (CEPP)
1986	Emergency Planning and Community Right-to-Know Act (EPCRA) * [also known as Superfund Amendment and Reauthorization Act Title III (SARA)] *Established LEPCs nationwide
1986	Chemical Accident Prevention Program
1986	Chemical Safety Audit Program
1987	Accidental Release Information Program (ARIP)
1990	Clean Air Act Amendments (CAA) section 112(r)
1992	**OSHA Process Safety Management (PSM) Regulation**
1996	**EPA Risk Management Program Rule (RMP)**

Pay special attention to the last two regulations in the list, OSHA PSM and EPA RMP. They represent a unique turn for regulatory agencies. Both were mandated to be developed under the Clean Air Act and can be considered unique due to the fact that they essentially require affected sites to establish a proactive business philosophy toward preventing and responding to chemical releases. They do this by requiring management systems to be developed. These management systems are typically written administrative procedures or policies describing how a site will comply with a key element of PSM and RMP. Neither regulation has many hard and fast detailed requirements. They primarily allow the affected site to develop its own unique methods for compliance with major elements and each requires the affected site to self-audit its performance on a regular basis.

For years OSHA and EPA regulations tended to be prescriptive, that is, they tended to be detailed and inflexible. They were written to solve one problem or set of problems often without leeway for individual situations or special considerations. The Process Safety Management standard and the Risk Management Program rule are examples of a new trend toward performance based regulation.

One other unique advantage these two regulations have is that they are based upon best practices that some companies and industry groups already recognized and had been developing and practicing on their own. The American Institute of Chemical Engineers (AIChE) established the Center for Chemical Process Safety (CCPS) in 1985. CCPS published a comprehensive process safety management tool in 1989 entitled *Guidelines for the Technical Management of Chemical Process Safety.* The Chemical Manufacturers Association (CMA) produced the *Responsible Care® Process Safety Code* in 1988 and by 1990 the American Petroleum Institute (API) had developed Recommended Practice 750, *Management of Process Hazards*, describing a process safety management system for its members.

Many companies had been using accepted hazard assessment techniques, risk modeling, and process safety management systems of their own accord for years prior to the passing of these regulations. Some of the more progressive companies recognized that these techniques not only enhanced safety and environmental responsibility but also allowed them to operate more efficiently and make high quality product. However, many other facilities could only see the costs involved with these practices and chose not to implement them, thus the regulations were born.

Although this book is focused on helping local emergency planning committees understand EPA's RMP rule, we reference

OSHA's Process Safety Management regulation in this book as it has essentially been adopted as a key part of the RMP rule termed "Prevention Program." OSHA PSM has been in force since 1992 and is generally well understood by facilities covered under it. We will compare the two regulations in more detail in Chapter 4 but let us look briefly at the purpose of each regulation to see how these two regulations combine:

Protecting the Public
Purpose of EPA's Risk Management Program
rule 40 CFR part 68

". . .to prevent accidental releases to the air and mitigate the consequences of such releases by focusing preventive measures on chemicals that pose the greatest risk to the public"

Protecting Employees in the Workplace
Purpose of OSHA's Process Safety Management Regulation
29 CFR 1910.119

". . .to prevent or minimize the consequences of catastrophic releases of toxic, reactive, flammable, or explosive chemicals"

Imagine that the EPA is like a sentinel standing with its back to the perimeter fence of a facility looking outward, guarding the public from accidental releases. OSHA is like a sentinel with its back to the inside of the perimeter fence looking inward, guarding the employees from catastrophic releases. In the case of these two regulations, RMP and PSM, both are intended to prevent chemical incidents. However, both regulations contain elements that show an understanding that even with a good prevention program, incidents may still occur and we must have a plan to address the immediate consequences and a system to prevent that particular failure (and ones like it) in the future. Our overall goal is to protect everyone and continuously improve our ability to do so.

Local emergency planning committees are key stakeholders in community risk management. They will be a primary end user of some of the data the EPA RMP rule requires facilities to prepare and submit. Each individual LEPC can choose to become more proactive in some areas as discussed in the next section.

What Can LEPCs Do to Enhance Public Safety Using the RMP Rule?

There are several activities that local emergency planning committees can undertake to assist in ensuring that the EPA RMP fulfills its purpose of protecting the public. These are

- Develop an understanding within the LEPC membership of the purpose and basic elements of the Risk Management Program rule.
- Work with facilities to help ensure their emergency response plans reflect and mesh seamlessly with the community emergency response plan.
- Actively evaluate additional information needs from the risk management plans and emergency response plans in your area and systematically seek out information from the stationary sources that will help your LEPC members respond effectively to the specific types of emergencies you may face.
- Act as a networking tool for the various stationary sources in your response area to assist in the transfer of information and compliance techniques between affected sites.
- Consider taking an active role in helping affected sites communicate their risk management plans to the public and in responding to requests from the public.
- Access and review the risk management plans for the facilities in your local community.
- Attend seminars provided for state emergency response committees (SERCs) and LEPCs on the implications of the EPA RMP rule. (CCPS offers public seminars for just this purpose.)

Each of these items is addressed in detail in the appropriate section of this book. The book as a whole is intended to help directly with the first item, developing an understanding of the rule and its elements. Many other resources are available to assist with overall knowledge of the RMP rule and its issues. A list of other references you can use is provided.

The next two items in the above list relating to the Emergency response plan are stated in the regulation itself and discussed in Chapter 3, the section entitled "Emergency Response Plan." EPA is explicit in stating that affected sites must ensure their emergency response plans complement your overall community plan and that affected sites must respond to LEPC questions concerning their plans.

The last two items in the list are not required by law but can be key to success for RMP implementation in your area. They are discussed in Chapter 5.

2

Developing Understanding: A Summary of the Risk Management Program Rule

Who Must Comply?

The facilities the RMP rule applies to are termed "stationary sources," as the rule does not apply to regulated substances under transportation.

A process that has materials from the EPA regulated substances list in quantities greater than the threshold quantity is a stationary source.

The term "process" means any activity where regulated substances are

- manufactured
- used
- stored
- handled, or
- transported on-site

A process includes vessels that are interconnected and separate vessels if located such that they could potentially be involved in a single release.

EPA estimates that approximately 66,000 facilities will be affected by the Risk Management Program rule. These can include

- chemical manufacturers
- petroleum refineries
- manufacturing facilities
- agricultural chemical retailers
- public drinking water and wastewater treatment systems
- utilities

- propane retailers
- cold storage facilities
- warehousing facilities
- military and energy installations

Take some time to list the names of facilities in your local community that you think may fall under the Risk Management Program rule. How many potentially affected sites are in your area of emergency planning responsibility? The actual number required to submit their risk management plan (RMPlan) to the Environmental Protection Agency (EPA) may be somewhat less than your estimate, as many affected sites are switching to nonregulated substances for certain processes or reducing inventories of regulated substances to below their threshold quantity. Remember, some facilities may fall under the RMP rule even if they did not fall under OSHA's Process Safety Management standard. For example, propane distributors are now covered under RMP.

Overview—Three Basic Parts

The EPA's Risk Management Program rule (40 CFR 68) requires stationary sources to implement a Risk Management Program and develop a risk management plan (RMPlan). "RMPlan" is a term coined by EPA and is short for risk management plan. It is a summary description of the risk management program activities carried out in the facility. A facility must submit its RMPlan to a central location from which the RMPlan will be available to regulators, local emergency planners, and the public.

A risk management program consists of three parts:

- a hazard assessment,
- a prevention program, and
- an emergency response program.

Successful compliance with the rule requires a stationary source to take on two additional tasks to ensure compliance initially and over time:

1. Establish an overall management system to develop and maintain the three components in an up-to-date condition.

2. Submit an RMPlan initially and resubmit updates as required by the rule.

Contents of the Risk Management Program Rule

The RMP rule was first published on June 20, 1996, in the *Federal Register* (FR). It contains both preamble language that explains EPA's goals for writing the rule and the regulatory text. EPA has published several guidance documents for companies that fall under the rule to use in their compliance efforts. Industry groups have also produced documents and model plans to assist their particular target audience in meeting the requirements of the rule. The American Petroleum Institute (API) has developed an example risk management plan for petroleum refineries and has also collaborated with the Chemical Manufacturers Association (CMA) to produce an overall RMP compliance guide that focuses on all provisions of the RMP rule. Stationary sources should consider using all of these resources when determining their coverage and when developing compliance strategies and implementation plans. Local emergency planning committees with stationary sources in their response area may find these additional resources useful as well.

The RMP rule has eight subparts, designated A through H. It also provides an appendix that lists toxic endpoints for use in hazard assessments. The titles of the subparts are listed below:

- Subpart A—General
- Subpart B—Hazard Assessment
- Subpart C—Program 2 Prevention Program
- Subpart D—Program 3 Prevention Program
- Subpart E—Emergency Response
- Subpart F—Regulated Substances for Accidental Release Prevention
- Subpart G—Risk Management Plan
- Subpart H—Other Requirements

A brief description of the contents within each subpart is given below. Refer to the text of the RMP rule in Appendix A for more detail.

SUBPART A—GENERAL

This segment describes the applicability requirements of the RMP rule. It sets the three-year compliance deadline; defines three different RMP program levels, including eligibility criteria and necessary work; and requires facilities to have a management system to control implementation of their risk management program. Program 1 is the least stringent RMP for "lower hazard" processes. A process qualifies for Program 1 if

- it has not had an accident with an off-site effect in the past five years,
- the worst-case scenario (WCS) endpoint distance determined by accepted modeling techniques does not reach the nearest public receptor of concern, and
- emergency response activities have been coordinated with local emergency planning committees.

A process is in Program 3 if it does not qualify for Program 1 and it is either

- covered by OSHA's Process Safety Management standard or
- considered within one of ten EPA selected North American Industrial Classification System (NAICS) codes. These are

NAICS Code	Industry
325181	Alkali and chlorine
325211	Plastics and resins
325311	Nitrogen fertilizers
32532	Pesticide and other agricultural chemicals
32411	Petroleum refineries
32211	Pulp mills only
32511	Petrochemical
325188	All other inorganic chemical manufacturing
325199	All other basic organic chemical manufacturing
325192	Other (covers cyclic crude and intermediate manufacturing)

If a covered process is not in Program 1 or Program 3, then it is eligible for Program 2, a streamlined version of Program 3 risk management. Page 17 at the end of this chapter contains a simplified flow chart that a facility can use to determine coverage under the rule and which program level they may be considered.

SUBPART B—HAZARD ASSESSMENT

This part of the rule consists of two parts:

- performing an off-site consequence analysis (OCA) of potential accidental releases and
- compiling a five-year history of accidental releases.

The off-site consequence analysis provides an estimate of the distance that a toxic vapor cloud or the effects from a fire or explosion could travel from a site. Two types of scenarios are defined. Worst-case scenarios (WCSs) and alternative release scenarios (ARSs). Definitions of WCS release conditions and modeling parameters are prescribed. Worst-case scenarios are modeled on incidents of extremely low probability. Alternative release scenarios, however, allow a stationary source to model a more likely incident. This is of particular interest when coordinating an emergency response plan with LEPCs as the alternative release scenarios become useful tools for emergency planning and provide realistic drill scenarios.

A facility must estimate the residential population within a circle defined by the distance calculated to the appropriate hazard endpoint centered at the assumed point of release. U.S. Census data are used to determine this number. If institutions, parks, recreational areas, major commercial areas, and sensitive environmental receptors are within the circle, they must be noted. Off-site consequence analysis data must be updated every five years, or more often if stationary source changes raise or lower the endpoint distance by a factor of two or greater.

SUBPART C—PROGRAM 2 PREVENTION PROGRAM

This segment details the prevention program requirements for Program 2 processes. The seven elements are

- Safety information

- Hazard review
- Operating procedures
- Training
- Maintenance
- Compliance audits
- Incident investigation

Each of these elements has distinct requirements. When compared to OSHA PSM counterparts however, they are not always as detailed.

SUBPART D—PROGRAM 3 PREVENTION PROGRAM

In this subpart, the prevention program requirements for Program 3 processes are specified, including the following twelve elements:

- Process safety information
- Process hazard analysis
- Operating procedures
- Training
- Mechanical integrity
- Management of change
- Pre-startup review
- Compliance audits
- Incident investigation
- Employee participation
- Hot work permit
- Contractors

The requirements in each element are essentially identical to their OSHA PSM counterparts. EPA has made a few terminology changes to ensure that facilities understand that they expect the prevention program to protect the public and the environment as well as workers.

SUBPART E—EMERGENCY RESPONSE

This segment of the rule describes emergency response requirements. Program 2 and 3 stationary sources whose employees are expected to respond to accidental releases of regulated substances must develop an emergency response plan designed to protect the public and the environment. Emergency response activities must also be coordinated with the community emergency planners and responders. Program 2 and 3 sites whose employees will not respond to accidental releases are not required to prepare an emergency response plan, but they must have a system to notify external emergency responders in the event of an accident. All covered facilities must respond to any requests from local emergency planners or responders for more information to assist with in preparing or revising the community emergency response plan.

SUBPART F—REGULATED SUBSTANCES FOR ACCIDENTAL RELEASE PREVENTION

This subpart contains the EPA list of regulated substances, their threshold quantities, and specific exemptions. The EPA list contains 77 toxic substances and 63 flammable substances. Except for one substance, methylchloride, the EPA threshold quantities are greater than the OSHA PSM threshold quantities. EPA describes a detailed method for determining whether a mixture of regulated and nonregulated substances is covered.

The RMP rule applies only to "stationary sources"; transportation activities such as pipelines and storage incident to transportation are currently not covered by the RMP rule. Additional exemptions to the rule are as follows:

- If a regulated substance is present in a concentration below 1% by weight of a mixture, the mixture need not be considered for determining a threshold quantity of the regulated substance.
- Gasoline, in distribution or related storage for use as a fuel, need not be considered when determining whether a threshold quantity of a substance is present at a stationary source.
- Naturally occurring hydrocarbon mixtures are exempted, including any combination of the following: condensate, crude oil, field gas, and produced water.
- Also exempted are exploration and production (E&P) facilities on the outer continental shelf (OCS)

SUBPART G—RISK MANAGEMENT PLAN

This segment details submitting the RMPlan, updating it, and the content requirements for an RMPlan. The RMPlan must contain three major sections:

1. an executive summary;
2. a certification that the information is true, accurate, and complete; and
3. a detailed list of data elements broken down into the following five categories:
   - Registration information
   - Off-site consequence analysis
   - Five-year accident history
   - Prevention program
   - Emergency response program

The original RMPlan for a stationary source must be submitted by the latest of the following dates:

- June 21, 1999;
- 3 years after the date on which a new regulated substance is listed; or
- the date on which a process is first covered.

EPA will use the postmark date on the submission as the official compliance date. It will be used to start the clock for the five-year update requirement. The RMP compliance center will track both postmark date and date received. In order to be in compliance, your submittal must be postmarked by June 21, 1999. If the postmark is illegible, the EPA will use the date received as the compliance date.

The RMPlan must be updated at least every five years, or within six months if specific changes occur that affect the basis of the RMP. EPA intends that facilities submit the RMPlan to a central point for access by regulators, local emergency planners, and the public.

An example RMPlan is provided in Appendix E of this text (see page 149).

SUBPART H—OTHER REQUIREMENTS

This subpart specifies the EPA recordkeeping requirements, availability of information to the public, the relationship between the RMP rule and air permits, and audits. Facilities must maintain RMP records for at least five years. The RMPlan is to be made available to the public, although any government classified information is protected by law.

If your facility holds a Title V Part 70 or 71 operating permit, the RMP rule is considered an "applicable requirement" under the permit. However, coverage under the RMP rule does not mean that you must necessarily obtain an air permit. It is important to note that the RMPlan is not a part of the air permit itself. Facilities with air permits must revise them to include either:

- a certification that a complete RMPlan has been submitted or
- a schedule for complying with RMP rule requirements.

The General Duty Clause

The "general duty" clause of the CAA Amendments of 1990 states the following:

> It shall be the objective of the regulations and programs authorized under this subsection to prevent the accidental release and to minimize the consequences of any such release of any substance listed pursuant to paragraph (3) or any other extremely hazardous substance. The owners and operators of stationary sources producing, processing, handling or storing such substances have a general duty in the same manner and to the same extent as section 654, title 29 of the United States Code, to identify hazards which may result from such releases using appropriate hazard assessment techniques, to design and maintain a safe facility taking such steps as are necessary to prevent releases, and to minimize the consequences of accidental releases which do occur.

The general duty clause applies to sites that contain RMP-regulated substances as well as sites that do not contain regulated substances (or do not contain regulated substances at or above the threshold quantity) but do contain other extremely hazardous substances (EHSs) not on the RMP list.

It is important to note that the definition of an extremely hazardous substance is not just limited to the EHSs listed in the Emergency Planning and Community Right-to-Know Act (EPCRA) Section 302. EPA is still determining how it will interpret and apply the general duty clause.

A facility that operates processes containing EHSs that are not considered regulated substances under RMP rule may wish to briefly summarize the risk management efforts applied to these processes. These risk management summaries could serve as tools for indicating compliance with the "general duty" requirements of the CAA and for communicating your risk management efforts to the public.

When does the general duty clause come into play? If a facility, not required to follow RMP due to having no regulated substances on site, releases a non–RMP-listed extremely hazardous substance, and there are off-site impacts, the company may be held to the RMP standards during EPA's investigation of the incident.

Program Level Selection

The flowchart on the facing page shows how a facility can determine coverage and program level.

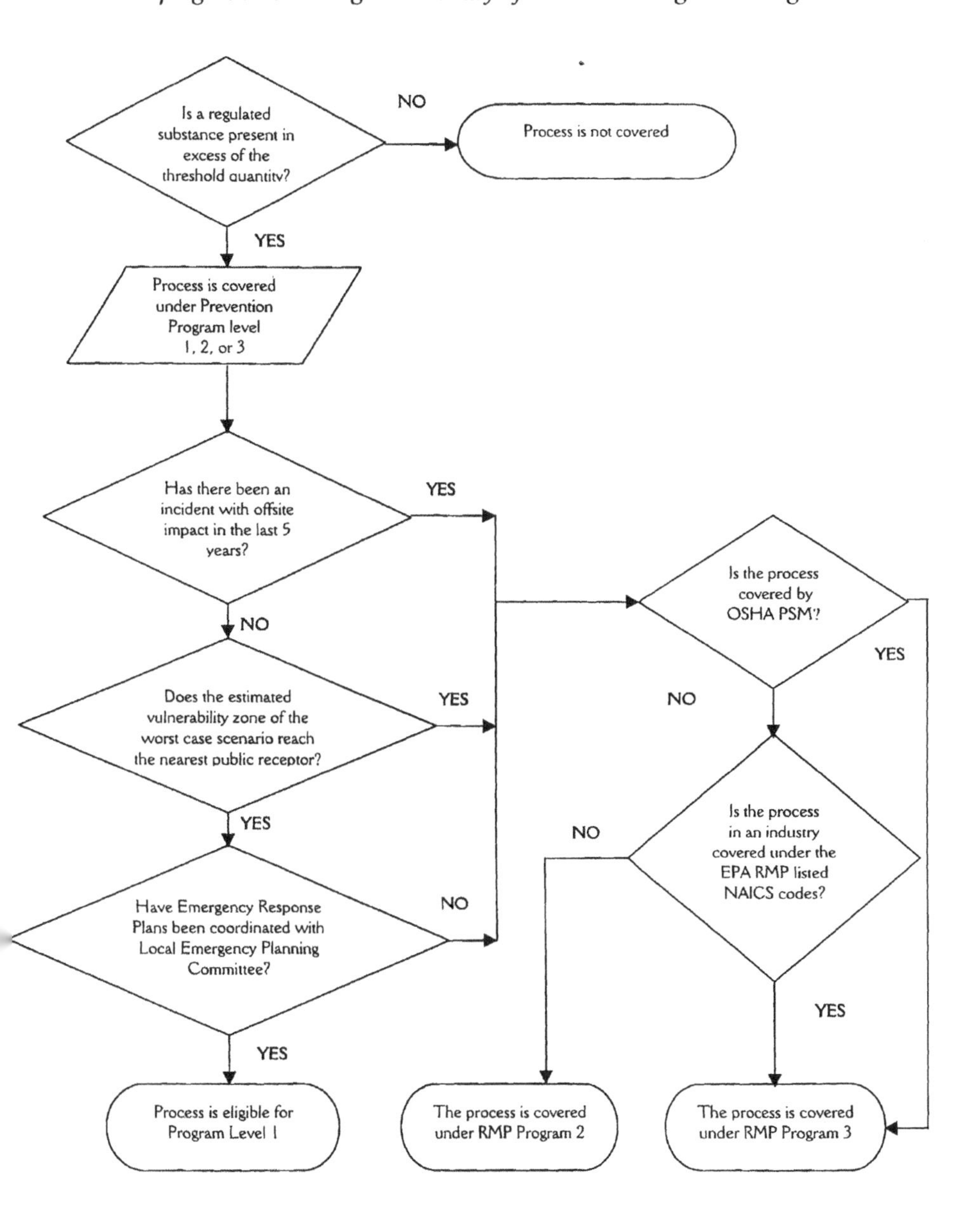
Is a regulated substance present in excess of the threshold quantity?
NO
Process is not covered
YES
Process is covered under Prevention Program level 1, 2, or 3
Has there been an incident with offsite impact in the last 5 years?
YES
NO
Does the estimated vulnerability zone of the worst case scenario reach the nearest public receptor?
YES
YES
Have Emergency Response Plans been coordinated with Local Emergency Planning Committee?
NO
YES
Process is eligible for Program Level 1
Is the process covered by OSHA PSM?
YES
NO
Is the process in an industry covered under the EPA RMP listed NAICS codes?
NO
YES
The process is covered under RMP Program 2
The process is covered under RMP Program 3

Program Requirements Summary

	PROGRAM LEVEL		
	1	2	3
Worst Case Release Scenario			
• One toxic or flammable for each Program 1 process	✓		
• Single toxic for all covered Program 2 or 3 processes (see note)		✓	✓
• Single flammable for all covered Program 2 or 3 processes (see note)		✓	✓
Alternative Release Scenario			
• At least one for each toxic in each covered Program 2 or 3 processes		✓	✓
• At least one to cover all flammables covered Program 2 or 3 processes		✓	✓
Five-Year Accident History	✓	✓	✓
Management System		✓	✓
Prevention Program		✓	✓
Emergency Response			
• Local agencies or facility provide; site must coordinate with response	✓		
• Develop and implement site program		✓	✓
Submission of RMPlan			
• Certification statement	✓	✓	✓
• Worst-case analysis results	✓	✓	✓
• Alternative case analysis results		✓	✓
• Five-year accident history	✓	✓	✓
• Data on prevention program elements		✓	✓
Note: Must submit additional worst-case scenarios for a hazard class if different public receptors are potentially affected.			

LEPC Quick Use Checklist

- ☐ **Use this chapter and Appendix A to become familiar with the basic parts of the RMP rule.**
- ☐ **Visit the EPA website at www.epa.gov/ceppo for more information and valuable guidance.**
- ☐ **Discuss the rule with other LEPC members and stationary source representatives.**

3

LEPC Use of Hazard Assessment Information

The hazard assessment portion of RMP is composed of two separate parts: the offsite consequence analysis and the five-year accident history. Together, they give EPA a snapshot of the potential for accidental releases, their severity, and a stationary source's past performance record in handling regulated substances.

Offsite Consequence Analysis

This segment of the rule requires stationary sources to perform analysis of two separate types of scenarios:

- worst case scenarios and
- alternative release scenarios.

The product of the analysis in both cases is a simple endpoint distance. This endpoint distance is used as the radius of a circle which, when placed on a map of the community and combined with population data, is used to determine what impact the modeled scenario will have on the public and the environment. One aspect needs to be pointed out about the resulting circles that show the endpoints for a site's off-site consequence analysis. Although the endpoint for an explosion can typically have a 360-degree impact that is not significantly impacted by wind speed and direction, toxic releases and fires (radiant heat flux) are greatly affected by wind speed and direction and rarely, if ever, impact through 360 degrees.

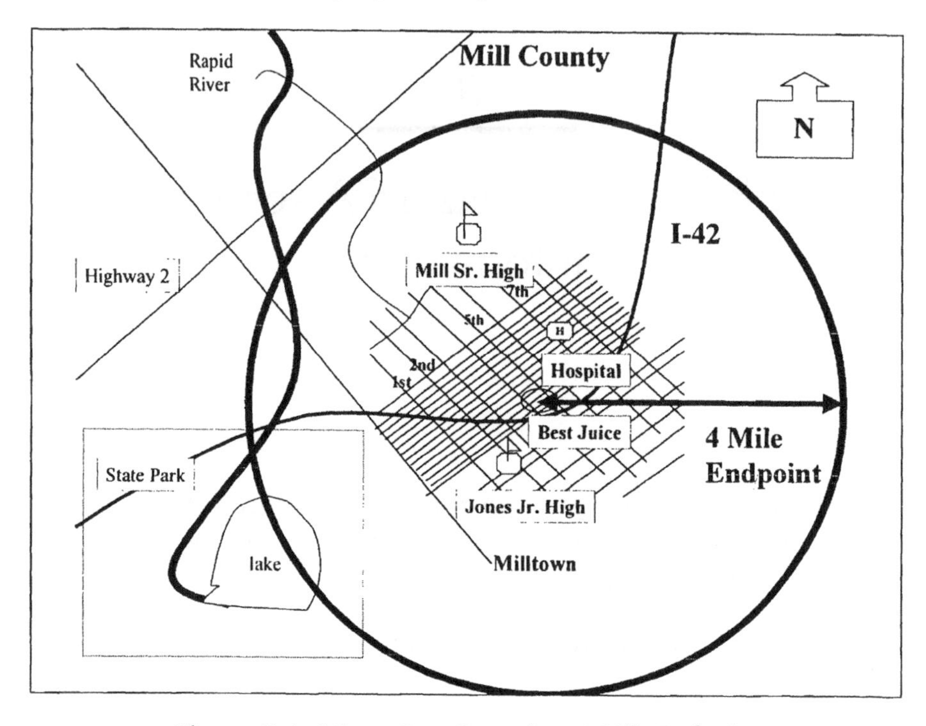

Figure 3.1. *Worst Case Scenario—4 Mile Endpoint*

Worst case release means the release of the largest quantity of a regulated substance from a vessel or process line failure that results in the greatest distance to a toxic or flammable endpoint.

Alternative scenarios are scenarios other than worst case. For alternative scenarios, sources may consider the effects of both passive and active mitigation systems. LEPCs should pay special attention to alternative release scenarios as they are based upon more probable events. These scenarios are not required for Program 1 processes.

The stationary source can use the lookup tables or use any technically valid modeling software or method to determine the endpoints for their scenarios. The difference between using the EPA lookup table and a commercially available modeling program is flexibility. The EPA table accepts only a few variables. A software-modeling program can accept specific variables for each type of release and a facility's location. The result is that the

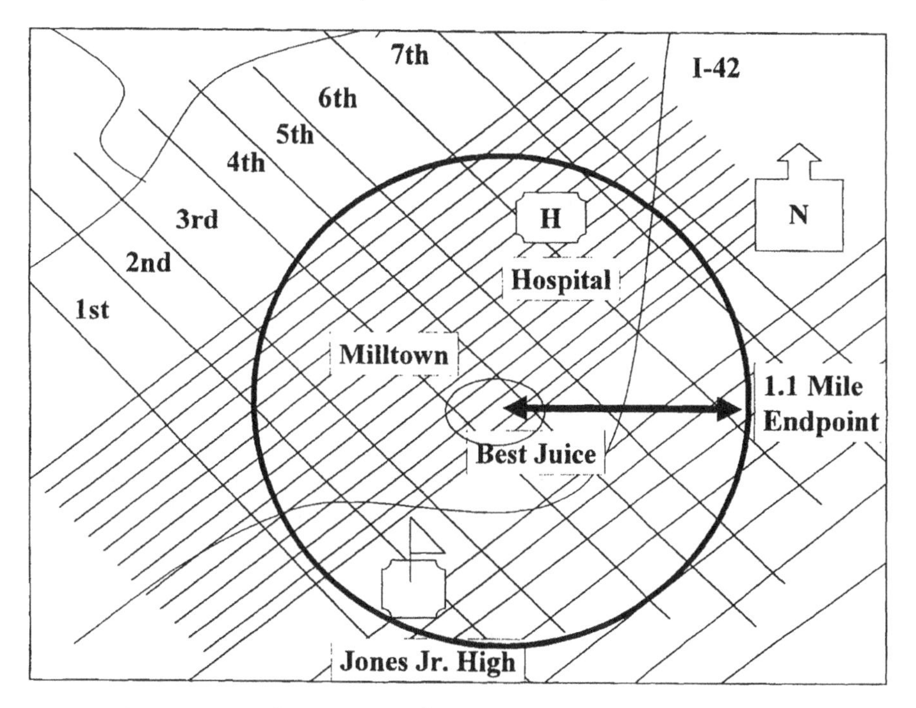

Figure 3.2. Alternative Release Scenario—1.1 Mile Endpoint

endpoint derived from the EPA table is usually more conservative, that is, a further distance, thus giving a larger potential impact area or *vulnerability zone*. Stationary sources that would like to reduce the distance to the endpoint may be tempted to use a custom modeling software program to show the smallest possible impact.

The approach some stationary sources are taking is to use the EPA lookup table for the worst-case scenario and a custom modeling program for the alternative release scenarios. One reason is that the worst-case scenario, as defined by EPA, is typically a very low probability event for a stationary source. It is wise to check the results for the alternative release scenario with both a modeling program and the lookup tables. A facility may find that the EPA lookup tables result in a suitable endpoint for both types of scenarios without having the potential burden of justifying the modeling system in the future.

Alternative releases on the other hand, are defined in a more flexible way. EPA allows companies to choose the type of release.

BASICS OF ALTERNATIVE RELEASE SCENARIOS

- They are not required for Program 1 stationary sources (lower risk users of regulated substances whose worst case scenario has no offsite impact)
- At least one must be developed for each toxic regulated substances above the threshold quantity at a stationary source
- At least one must be developed for one representative flammable regulated substance above the threshold quantity at a stationary source
- No set scenarios mandated—facilities have much more latitude in defining a realistic case

Sites can consider their accident history, the hazard reviews conducted for their process, or look at their industry's history for their process to select an alternative release scenario. Some situations you may see described in their RMPlans are

- Transfer hose releases due to splits or uncouplings
- Process piping releases from failures of flanges, welds, valves or drains
- Process vessel or pump failures due to cracks, seal failures, and open bleed valves
- Vessels overfilling
- Vessel or piping overpressurization and relief
- Mishandling of shipping containers

What Do We Do with Offsite Consequence Analysis Data?

LEPCs gain the most useful data by questioning stationary sources in their community about the specifics of their alternative release scenarios. These are more realistic events than worst case scenarios and could possibly be seen at some time during a facility's life span.

Several uses come to mind:

- Use alternative release scenarios to build decision trees as when to implement shelter-in-place or evacuation for residents near each stationary source in your response area.

- Use the alternative release scenarios to plan out specific public evacuation routes and sequences around a given stationary source.
- When planning for drills and exercises, use actual alternative releases as the scenario you are responding to. The staff at stationary sources will gain as much as your member organizations by mocking-up these types of scenarios.

RMPlan data required for OCAs is limited. However, responsible and forward thinking companies will be more than willing to work with your committee to plan a thorough emergency response to their alternative release scenarios.

Current thinking is that the EPA's intent for WCS data is as a quick screening tool to classify stationary sources by their potential impact on the public and the environment. Worst case scenario data allow LEPCs to pinpoint potential response areas and maximum vulnerability zones for their response area. But it is somewhat daunting to map out all the WCS vulnerability zones for any locality with multiple stationary sources. The typically long endpoints for common regulated substances such as chlorine can be interpreted as just too much for any emergency response organization to effectively deal with. In reality, a worst case release would most probably not affect the full circular area required to be reported by EPA. It would be a rare situation indeed for a worst case scenario toxic release to occur at the same time the wind direction shifted through 360 degrees to provide a perfectly consistent deposition to the endpoint. Again, WCSs are screening tools for overall risk.

LEPC Quick Use Checklist

- ❑ **ARSs are useful for building decision trees for determining when to call for public shelter-in-place or evacuation.**
- ❑ **WCSs are quick pinpoints of potential problem areas.**
- ❑ **ARSs are useful for emergency planning of evacuation routes.**
- ❑ **ARSs are useful for planning drill scenarios and training exercises.**

Five-Year Accident History

The RMPlan data for each stationary source's five-year accident history is intended as a measure of how well a stationary source has performed in regard to accidental release prevention during a specific five-year period. The information it provides is fairly simple. The wording from the rule is paraphrased below:

The five-year accident history shall include all accidental releases (of regulated substances) from covered processes that resulted in

- onsite deaths,
- onsite injuries, or
- significant onsite property damage

or known

- offsite deaths,
- offsite injuries,
- offsite evacuations,
- offsite sheltering in place,
- offsite property damage, or
- offsite environmental damage.

The owner or operator shall report the following information for each accidental release, included in the history (numerical estimates may be provided to two significant digits):

- date, time, and approximate duration of the release;
- chemical(s) released;
- estimated quantity released in pounds;
- the type of release event and its source;
- weather conditions, if known;
- on-site impacts;
- known offsite impacts;
- initiating event and contributing factors if known;
- whether offsite responders were notified if known; and
- operational or process changes that resulted from investigation of the release.

What Do We Do with Five-Year Accident History Data?

Fortunately, many stationary sources will be able to state in their RMPlan that there were no onsite or offsite accidental releases that met the requirements for reporting. However, for those facilities in your response area that must report on or off site releases, consider planning to drill on responses to that specific type of release. The information in the RMPlan plus the company's incident investigation records could combine to produce an interesting and useful tabletop analysis of successful response and areas for improvement.

An example RMPlan is included in the Appendix E. Take a look at the type of data reported.

LEPC Quick Use Checklist

- ❑ **Five-year accident histories are useful for planning realistic drill scenarios**
- ❑ **Five-year accident histories are useful for tabletop study to help understand the best practices and weaknesses from past performance.**
- ❑ **They are also useful for evaluating process safety performance at the stationary source.**

What Do We Do with Five-Year Accident History Data?

[illegible]

4

LEPC Use of RMP Emergency Response Plan Information

The section of the rule governing the emergency response plan specifically mentions local emergency planning committee involvement. It makes it clear to stationary sources that the LEPC must have their questions answered in regard to the facility's emergency response plan.

The EPA had four intentions for this portion of the rule:

1. To identify sources of substances that could affect the public or the environment.
2. To reduce the severity of exposures to regulated substances through proper response.
3. To provide timely warning to members of the public that could be affected by a release.
4. To coordinate information on specific source emergency response to local emergency response agencies.

What Must Facilities Do to Comply?

If a facility's management is organized so that employees **do not** respond to an accidental release, then the following conditions must be met:

- Toxic regulated substance release response must be included in the community emergency response plan.
- Flammable regulated substance release response must be coordinated with the local fire department.
- A mechanism must be in place to notify emergency responders.

If a facility's management is organized so that trained employees **do** respond to an accidental release, then the wording from the rule paraphrased below applies:

- The owner or operator shall develop and implement an emergency response program for the purpose of protecting public health and the environment. Such program shall include the following elements:
  - —An emergency response plan, which shall be maintained at the stationary source and contain at least the following elements:
    - » procedures for informing the public and local emergency response agencies about accidental releases;
    - » documentation of proper first-aid and emergency medical treatment necessary to treat accidental human exposures; and
    - » procedures and measures for emergency response after an accidental release of a regulated substance;
  - —Procedures for the use of emergency response equipment and for its inspection, testing, and maintenance;
  - —Training for all employees in relevant procedures; and
  - —Procedures to review and update, as appropriate, the emergency response plan to reflect changes at the stationary source and ensure that employees are informed of changes.
- A written plan that complies with other federal contingency plan regulations or is consistent with the approach in the National Response Team's Integrated Contingency Plan Guidance ("One Plan") and that includes other elements provided this section shall satisfy the requirements of this section if the owner or operator also complies with the next paragraph.
- The emergency response plan shall be coordinated with the community emergency response plan developed under 42 U.S.C. 11003.
  - —Upon request of the local emergency planning committee or emergency response officials, the owner or operator shall promptly provide to the local emergency response officials information necessary for developing and implementing the community emergency response plan.

Many facilities are considering revising their emergency response programs to match the intent of the NRT's Integrated Contingency Plan approach (ICP) mentioned in the regulation. The ICP or "one-plan" is a way to avoid the trouble of maintaining multiple documents to meet overlapping laws such as

- EPA's Oil Pollution Prevention Regulation (Spill Prevention Control and Countermeasures—SPCC and Facility Response Plan)—40 CFR part 112.7 (d) and 112.20-21
- EPA's Risk Management Program Rule—40 CFR part 68
- EPA's Resource Conservation and Recovery Act—Contingency Planning Requirements—40 CFR part 264, subpart D, 40 CFR part 265 subpart D, and 40 CFR part 279.52
- OSHA Process Safety Management of Highly Hazardous Chemical Substances Regulation—29 CFR 1910.119
- OSHA's Emergency Action Plan Regulation—29 CFR 1910.38 (a)
- OSHA's Hazardous Waste Operations and Emergency Response (HAZWOPER) regulation 29 CFR 1910.120
- Other state or federal regulations that may affect a site's emergency response.

This approach simplifies the paperwork and training associated with emergency response plans and gives LEPCs a central location for finding all the specific emergency response activities for a given site.

What Is the LEPC Connection to RMP Emergency Response Plan Requirements?

The sentence from the rule repeated below is what makes your LEPC a key stakeholder in the results of the RMP rule.

> Upon request of the local emergency planning committee or emergency response officials, the owner or operator (of the stationary source) shall promptly provide to the local emergency response officials information necessary for developing and implementing the community emergency response plan.

LEPCs become a more proactive stakeholder as EPA is inviting your committee to ask more questions of the facilities around you. The specific methods for delivery and access to RMPlan information for stationary sources in your response area is still under development at the time of this writing. However, it is clear at that LEPCs will be given a "golden password" to the database that contains the full RMPlan information.

It is up to you to use this database to come up with questions that will help your LEPC revise as necessary and implement a community emergency response plan that takes into account the potential effects of the regulated substances in your area.

Please note that LEPCs are not limited in any way as to what information they may request. If you feel it is essential to know specific information on the modeling program a company used for its alternative release scenarios, you can ask. If you would like to see a company's entire emergency response plan, just ask. If you would like to see their entire prevention program, just ask. Written requests are recommended for documentation in the event a facility hesitates to respond.

The nationwide RMP registration effort implies that facilities will be asking for copies of your community emergency response plan to ensure theirs will mesh. The information LEPCs obtain from the emergency response plans produced for RMP compliance should be used to revise and improve your local plan.

Some specific requests for information that may help you plan for this phase of RMP implementation and improve your community emergency response plan follow:

- Ask for a current copy of a facility's emergency response plan and a chance to discuss questions with its "owner."
- Request the specific modeling information for the alternative release scenarios and an opportunity to discuss how it was chosen with the person responsible for RMP compliance.
- Consider asking for the administrative documents that define a company's prevention program or process safety management (PSM) system. These will give you a feel for the site's grasp and control of accidental release prevention methods and systems.

As you review the information, look at the various formats and techniques the stationary sources in your area use in their plans. Are there any ideas in their plans which you could use to upgrade your community emergency response plan?

LEPC Quick Use Checklist

- ❑ **Once RMP data are made available by the EPA, learn to use the software to search and scan the RMPlans for the facilities in your community.**
- ❑ **Systematically begin requesting emergency response plans for review and arrange meetings to discuss them.**
- ❑ **Target those facilities whose RMPlans indicate more attention and request further information such as detailed alternative release scenarios and prevention program data.**
- ❑ **Approach facilities for assistance in planning out drill scenarios using actual data from their RMPlans.**

5

LEPC Use of Prevention Program Information

A stationary source under program level 2 or 3 is required to have a prevention program. Of the three parts of the Risk Management Program rule; hazard assessment, emergency response plan, and prevention program, most persons agree that the prevention program is the most important piece in avoiding accidental releases and protecting the public and environment.

A high-quality prevention program that is maintained and made a key part of a facility's management philosophy can help us avoid ever realizing the hazards of accidental releases. The requirements for a prevention program are different for each program level. These requirements are as follows:

- The RMP rule does not require a formal prevention program for Program 1 processes.
- Subpart C defines Program 2 prevention program requirements.
- Subpart D defines Program 3 prevention program requirements.

The Clean Air Act [§112(r)] requires the Risk Management Program to include a prevention program that addresses safety precautions and maintenance, monitoring, and employee training.

The RMP Program 3 prevention program regulatory language is nearly identical to the OSHA Process Safety Management standard with minor wording changes to address the differences. Some items, such as completion dates and process hazard analysis revalidation efforts are coordinated with OSHA Process Safety Management requirements.

The RMP Program 2 requirements include most of the process safety management elements as the basis for sound prevention practices, but it is designed for less complex processes. A stationary source eligible for Program 2 has a reduced documentation and recordkeeping burden.

The Relationship between EPA/RMP and OSHA PSM

Chapter 1 presented and compared the stated purposes of EPA RMP and OSHA PSM. In summary; the Process Safety Management regulation is protecting the workers inside the fence from hazardous materials, the Risk Management Program rule is protecting the public and the environment outside the fence from hazardous materials. Both use some of the same tools to do this. The RMP rule specifies that a company that currently falls under OSHA PSM satisfies the requirement for a prevention program ***if*** their PSM system is in compliance. As described in chapter 1, industry was establishing management systems to help control risks and the potential effects on the public several years before the EPA RMP rule and OSHA PSM regulation were ever in force. Table 5-1 shows the similarities and differences of the major programs.

LEPC Interaction—Prevention Program

Reviewing the prevention program for a facility may provide a clearer insight to a stationary source's philosophy of risk management. Though not essential to fulfilling the stated portion of the RMP rule that requires stationary sources to respond to LEPC questions, the prevention program can tell you a great deal about the importance a facility places on release prevention. The text of the OSHA Process Safety Management regulation and a detailed description of each element is included in the appendix. The EPA RMP wording is also available for comparison. Here is a quick summary of the intent of each element of a prevention program.

TABLE 5-1
A Comparison of RMP Elements and Various Process Safety Management Systems

EPA RMP Implementation Items	OSHA PSM	API RP-750	CMA Responsible Care© RP-9000 (STEP)	Process Safety Code	CAER Code
Registration					
Management System					
Hazard Assessment					
Offsite Consequence Analysis					
Accident History				✓	
Prevention Program					
Process safety information	✓	✓	✓	✓	
Process hazard analysis	✓	✓	✓	✓	
Operating procedures	✓	✓	✓	✓	
Training	✓	✓		✓	
Mechanical integrity	✓	✓	✓	✓	
Management of change	✓	✓	✓	✓	
Pre-startup review	✓	✓		✓	
Compliance audits	✓	✓		✓	
Incident investigation	✓	✓	✓	✓	
Employee participation	✓	✓	✓	✓	
Hot work permit	✓			✓	
Contractors	✓		✓	✓	
Emergency Response Program	✓	✓	✓		✓
Risk Management Plan					
Communications to the Public			✓	✓	✓

PROCESS SAFETY INFORMATION (PROGRAM 3 AND PROGRAM 2)

EPA and OSHA are saying to stationary sources, "know your processes." This element asks facilities to document information about the equipment, materials of construction, chemicals, and reactions involved in what they do.

PROCESS HAZARD ANALYSIS (PROGRAM 3 AND PROGRAM 2)

This element of a prevention program asks facilities to use an accepted, systematic method for evaluating the hazards of operating the processes they use. It asks them to look at design and equipment (based upon their process safety information) as well as human factors and facility siting to determine whether physical or administrative changes can be made to reduce the hazards of operating a process.

OPERATING PROCEDURES (PROGRAM 3 AND PROGRAM 2)

Prior to OSHA's PSM regulation, it was not uncommon to find a chemical processing facility operating with outdated (or nonexistent) operating procedures. This element asks facilities to document the steps needed to perform safe startup, normal operation, shutdown, and emergency shutdown for the processes at the stationary source. They must be up-to-date and accessible to employees.

TRAINING (PROGRAM 3 AND PROGRAM 2)

This element requires facilities to provide three types of training:

- *process overview training* on how the whole system works at their location and the processes unique hazards,
- *procedure training* on the specific operating procedures an employee is expected to perform, and
- *refresher training* on any process information or procedure information that changes. Refresher training is required at least every three years or more often as requested by employees.

MECHANICAL INTEGRITY (PROGRAM 3 AND PROGRAM 2)

This element asks facilities to inspect and test the physical equipment that makes up their process. It requires that maintenance procedures exist and that employees are trained on the maintenance procedures and an overview of the process. Facilities must maintain records of repairs and other activities related to mechanical integrity.

MANAGEMENT OF CHANGE (PROGRAM 3)

OSHA and EPA want facilities handling hazardous chemicals to control the changes they make to the materials, equipment and operating methods used in the processes. Prior to industry adopting a process safety management philosophy, changes were often made without complete documentation or a thorough study on the change's impact to health, safety, and the interconnected processes and equipment. This element simply says, if you are going to make a change to your process that is not a direct "replacement in kind," you must look at specific process safety items related to the change.

PRE-STARTUP REVIEW (PROGRAM 3)

This element establishes a formal good practice many companies and facilities have used for years. It asks that PSM and RMP covered processes receive a special review for any major modifications or plant additions prior to introducing hazardous chemicals or regulated substances. It is a second-level check of the management of change process to ensure training is complete, procedures are written , equipment is installed according to the design specifications, and the process is safe for startup.

COMPLIANCE AUDITS (PROGRAM 3 AND PROGRAM 2)

This is a key element to process safety management and any risk management program. It requires covered facilities to perform a self-audit of their own program on a regular basis. This makes a facility responsible for knowing the status of its risk management and process safety management program and taking required actions to continuously improve them.

INCIDENT INVESTIGATION (PROGRAM 3 AND PROGRAM 2)

Like pre-startup safety review, this element establishes a requirement for a good practice industry has been using for years. Whenever an incident occurs, an investigation must be performed to log the specific details of the incident, examine its causes, and determine actions to take to prevent further recurrence of the incident. RMP requirements add a need to tie the results of the incident investigation to the five-year accident history to ensure stationary sources maintain proper records and updates for their RMPlan.

EMPLOYEE PARTICIPATION (PROGRAM 3)

This element requires facilities to document the methods they use to involve employees in the process hazard analyses and in all the other elements of a process safety management program or risk management Program 3 prevention program. OSHA and EPA both realize that employees are often very knowledgeable about the equipment, maintenance activities, and operating practices. It is essential that the employees are involved in the implementation of any overall management system designed to protect them and the public from the inherent risks resident in the activities they are trained for and paid to perform on a daily basis.

HOT WORK PERMIT (PROGRAM 3)

This element reemphasizes an existing OSHA standard that requires sites to have a way to govern the use of welding, burning, and cutting activities that occur near processes containing regulated substances or hazardous chemicals.

CONTRACTORS (PROGRAM 3)

In this element, OSHA and EPA require facilities to select contractors that have proven records of safety in regard to their people and their business. It delineates responsibility for both the owner/operator and the contract owner or operator. It also asks companies to verify that contract personnel receive specific training related to the process chemicals and equipment they are assigned to work with.

6

Proactive Aspects of LEPC Involvement in Risk Management

This chapter addresses the things your LEPC might do to show a strong philosophy of excellence as a proactive stakeholder in the risk management activities for your response area. In the best case, some stationary sources in your community have already approached your LEPC to invite participation in information sharing and communication activities. Below is a paraphrased version of a letter developed by a team of plants that were seeking to involve their LEPC in risk management activities.

TO: Milltown LEPC Chairman

FROM: Milltown Area Chemical Manufacturers

SUBJECT: *Request for the LEPC to develop a Risk Management Plan Subcommittee*

Dear Fire Chief Smith:

As you know, the EPA Risk Management Program rule requires all affected stationary sources to submit a Risk Management Plan (RMPlan). The Milltown Area Manufacturers propose the LEPC create a subcommittee of representatives from all stationary sources within Mill county that includes those affected by Risk Management Program rule requirements as well as all appropriate area emergency responders.

The Milltown Area Manufacturers will assist this subcommittee in providing direction for the development of Risk Management Plans (RMPlans) which will be useful to the LEPC. We envision an LEPC subcommittee with a charter to

- Provide guidance to assist all local stationary sources to consistently comply with Risk Management Program rule requirements.
- Oversee public education and awareness for the information contained within all RMPlans.
- Develop LEPC responses for public information requests regarding RMPlans.

The Milltown Area Manufacturers are willing to provide the LEPC with manpower and resources to help achieve these goals. Some additional benefits this subcommittee may realize include

- Providing the LEPC with useful information in the RMPlans to evaluate the current Mill County Emergency Response Plan and upgrade it if needed.
- Conducting local media campaigns to enhance public awareness of emergency response including shelter-in-place and other techniques.
- Assisting municipal facilities and smaller enterprises in their enhancement of public safety and environmental responsibility through their RMPlans.
- Increasing public awareness of the LEPC's presence and mission.

Our local LEPC has a reputation for being one of the country's most proactive committees. Several valuable ideas have been identified to help maintain this status in regard to the Risk Management Program rule.

A stationary source's minimum legal requirement is to simply submit a Risk Management Plan to the EPA. EPA will be making the RMPlan information available to the public in a format and level of detail the agency feels is appropriate. However, being responsible industry leaders, we feel it is important to go further in educating the public regarding the content of this rule and in how individuals can use this information to increase their knowledge and safety regarding the risks of potentially hazardous operations.

Our representative will be contacting you within the next week to discuss the possibility of forming this subcommittee in more detail. Our team members will represent their respective plants and act in our behalf in regard to this effort. We thank you in advance for your consideration and look forward to working with you on this issue.

Sincerely,

Milltown Chemical Plant Managers

The following example letter was later sent from the Subcommittee to local facilities and businesses identified from the SARA title III EHS registrations the LEPC received. It was intended to bring in some facilities, such as propane distributors, that may not realize they are covered under RMP.

TO: All Processors and Handlers
of Hazardous Materials

SUBJECT: *Invitation to Participate in the Mill County LEPC Risk Management Program Subcommittee*

As you may know, EPA's Risk Management Program rule requires all affected facilities to develop and submit a Risk Management Plan (RMPlan) by June 21, 1999. The Mill County Local Emergency Planning Committee (LEPC) has created a subcommittee to assist organizations in determining their status as stationary sources and in developing an RMPlan. Our goal is to help local stationary sources comply with RMP requirements and integrate RMPlans into the Mill County Hazardous Materials Plan.

The subcommittee is comprised of stationary sources and appropriate emergency responders that are working together to develop

- **A Hazard Assessment Protocol:** This is a guide to use when completing the hazard assessment portion of the RMPlan. Use of the hazard assessment protocol will result in consistent results throughout Mill County.

- **A Communication Schedule:** A schedule of communication activities involving a range of groups from source employees to the public. The communication schedule will allow stationary sources to communicate similar information at the same time.
- **A network to Share Experience and Technical Issues:** The subcommittee is made up of technical personnel with a good working knowledge of the regulation. This allows for quick answers to technical questions concerning the regulation requirements.

Participation is essential to this subcommittee's success. If all local stationary sources become involved, we can achieve a positive impact on public safety.

The subcommittee mission statement and the meeting minutes from the initial subcommittee meeting are attached. Meetings are held once a month immediately following the LEPC meeting. The next meeting will be held on May 14 at 2:30 PM at the Emergency Operations Center.

If your company is covered by this regulation, if you anticipate being covered in the future, or if you are not sure about the applicability of this regulation to your facility, please attend the next meeting. If you have questions or comments you can contact me at 555-1212.

Sincerely,

Director, Emergency Services, Mill County
Chairman, LEPC RMP Subcommittee

In this case, the stationary sources were proactive enough to recommend this team oriented approach to their LEPC. You may choose to modify these two letters and send them to the potential stationary sources in your community to encourage them to work together. Discussion follows on the benefits and methods you might choose to seek for your community.

Assisting Your Community's Stationary Sources in Complying

If your response area has several stationary sources that are struggling with RMP compliance, your organization has a chance to bring these stakeholders together to share knowledge and resources to reach a common goal. Smaller facilities and some municipal stationary sources just don't have the manpower or resources to easily tackle the modeling, management system development, and recordkeeping required for RMP compliance.

Your LEPC can play a role in bringing these smaller enterprises up to speed in knowledge of the RMP rule and the techniques and resources available to assist in compliance. One tool that was a success in the East Harris County Manufacturers Association (EHCMA) RMP efforts in the Houston Ship Channel area of Texas was the development of a hazard assessment protocol to help companies in an industrialized area be consistent in the modeling techniques and alternative release scenario selection. This can also be a great help to the LEPC as these are the portions of the rule that can affect revision of the community emergency response plan. Offsite consequence analysis for worst case and alternative release scenarios are the more technically demanding activities that a stationary source must address, and smaller players would appreciate some guidance and support.

The Mill county plants that banded together to support their LEPC realized that, although they may sometimes be competitors in business, they were teammates in complying with RMP and in their communities' emergency response readiness.

Another aspect of the RMP rule is that the information in the RMPlan be made available to the public. The EPA is debating the specific methods for doing this but what is known is that the minimum legal requirement for a facility is to register their RMPlan with the EPA. The EPA will then make it available via local reading rooms, electronic files, or internet access. Various levels of access are being considered as well.

Though a stationary source meets this portion of the rule by registering, many sites are choosing to be more open in describing their RMPlans to the community to enhance their understanding of the risks involved in a given process. Reasoning for doing this is backed by the potential for misunderstanding the meaning of the worst case scenario vulnerability zone. It would be easy for a person with a non–chemical industry background to be amazed at the number of persons potentially affected by a worst case scenario by something as common as their local waste treatment facility. By using the LEPC to educate the public, some localities are seeking to belay this effect.

The LEPC targeted by the Mill County Chemical manufacturers became interested in playing a role in helping the local government and stationary sources in developing their understanding of the RMP rule. They contacted their EPA regional office and arranged for the EPA's local expert in RMP to come to their city to present a full day's seminar on RMP compliance. The meeting was well attended.

LEPC Quick Use Checklist

- ❑ **Prepare a letter to send to all potential stationary sources inviting them to participate in forming a subcommittee to address RMP compliance.**
- ❑ **Contact your regional EPA office to see about nearby presentations of their RMP compliance seminar.**
- ❑ **Consider developing a Hazard Assessment protocol document with the stationary sources in your area to ensure consistency in data related to accident history and offsite consequence analysis.**

Acting as a Clearinghouse for Community RMPlan Information

If a team effort is organized for the stationary sources in your community to comply with RMP, an opportunity will arise for the LEPC to be the central clearinghouse for information related to Risk Management in your area. A media campaign may be appropriate prior to the registration date to establish a central LEPC telephone number or website address for the public to access to learn the basics of Risk Management Plans for the stationary sources in their community.

If there exist community action groups in your area that target the environmental performance of local industry, the LEPC could act as the neutral territory for hosting open forums where facility stationary source representatives assemble to field specific questions from the public. The LEPC's presence would be essential in order to respond to questions related to readiness and manpower.

If your LEPC hosts a website, it can serve as a bulletin board for updates on RMP activities and, as we shall see in the next section, basic information useful for responding to any kind of community emergency.

Ensuring the Public Understands Its Role in Risk Management

Totally separate from RMP, but as a direct result of government's focus on risk management, is the emphasis on the goal of ensuring that every person in the LEPC's service area understands the techniques that make up efficient sheltering-in-place.

The RMP effort will bring the stationary sources in close communication with your LEPC at least for a time period around their registration, and hopefully longer, to sustain community emergency preparedness.

This is an excellent time to renew your campaign to ensure public awareness of how to behave during any emergency.

One proven technique that every member of the public needs to know is how to shelter-in-place and await further instructions. For emergencies involving regulated substances, sheltering in place can prevent unnecessary impact on the public from alternative release scenarios modeled for RMP compliance and the highly improbable worst case scenario.

A few specific techniques that have been used by LEPC's around the country are listed below:

- Engage local grade-school or middle-school-age students in a contest to illustrate shelter-in-place techniques. Display their drawings with text for the techniques on posters, your website, or in a local public service announcement (PSA) on television or in area newspapers. The Corpus Christi, Texas LEPC used this technique and posted it on a website.

- Have a page of the local telephone directory devoted to emergency response activities, phone numbers and shelter-in-place techniques.
- Prepare a brochure for the LEPC stating its overall function and phone numbers along with shelter-in-place methods. The brochure should be provided to newcomers to the area as a part of the local "welcome wagon" or chamber of commerce welcoming activities.
- Develop or revive a "mascot" program such as the *Wally-wise* shelter-in-place turtle character (Corpus Christi, Texas LEPC) or the *Sammy Siren* character (Billings, Montana LEPC) designed to build awareness in the area. Arrange for appearances at community gathering places.
- Establish (or include) in an LEPC website the steps for efficient shelter-in-place techniques and the media contacts to make for further information.

Suggested Activities

A checklist is provided below to help you determine what activities this concept book has presented will be most helpful to your community. They may not all be appropriate for you to attempt nor may they all be achievable in your community. However, they could be a starting point or a progress check for your community risk management efforts. Some stationary sources in your community may be agreeable to assisting with resources to achieve some activities.

DEVELOPING UNDERSTANDING

- ❑ Get your LEPC members to review the RMP rule and research and discuss the rule by reading this book or other resources.
- ❑ Have outside speakers present the basics of RMP at an LEPC meeting. They could be from the EPA, state agencies or local stationary sources.

Personal Safety Plan

Shelter-in-Place

Shelter-in-place should be taken seriously. Here are some safety steps to follow:

1. Quickly get into your house or shelter.
2. Shut all doors and windows.
3. Turn off all heating and cooling systems.
4. Tape and seal all windows and doors.
5. Place wet clean towels under doors to absorb gases.
6. Keep everyone in a room that has the fewest windows and doors.
7. Stay off the phone to keep lines of communication open.
8. Listen to TV or AM 620 radio for further instructions.

Shelter-in-Place Kit

To make your own SHELTER-IN-PLACE kit include the following:

1. First Aid Kit
2. 2 days worth of non-perishable items
3. Duct tape
4. Clear plastic to cover windows
5. Towels
6. Flashlight and extra batteries
7. Radio and extra batteries
8. Two gallons drinking water

Special thanks to the students of Tuloso-Midway Intermediate School for the Shelter-in-Place artwork.

Figure 6.1. *Shelter-in-Place Excerpt from the Nueces County LEPC Website. Reprinted with permission of the Nueces County LEPC.*

- ❑ Attend a seminar on RMP compliance or arrange for one in your community.
- ❑ Develop one or more persons in your group as experts to answer RMP related questions from the public. Have them coordinate with a representative from each stationary source.
- ❑ Identify regulated facilities in your area by using existing EPCRA information or actively inquiring through written or telephone contacts with facilities.

ASSISTING STATIONARY SOURCES IN COMPLIANCE

- ❑ Form a subcommittee or more informal group of LEPC members and representatives from all the stationary sources in your area to discuss successes and problems in their attempts to comply with RMP.
- ❑ Develop a hazard assessment protocol document to ensure consistency in OCA data and techniques.
- ❑ Become a resource to help the facilities in your area find other resources (state, associations, EPA) that can assist.

ENHANCING PUBLIC KNOWLEDGE

- ❑ Coordinate an RMP project for your area for informing the public and the media about the rule and its impact on the community (for example, like Kanawha Valley West Virginia and Augusta/Richmond County, Georgia).
- ❑ Perform live public outreach activities for shelter-in-place activities to raise awareness.
- ❑ Place advertisements in print media, radio, or television (or all three) listing the steps for sheltering-in-place techniques.
- ❑ Develop a brochure on the basics of the RMP rule.
- ❑ Develop a brochure on shelter-in-place techniques for distribution.
- ❑ Consider a website that can be used to provide shelter-in-place information and the basics of the RMP rule. If e-mail can be sent through the website, it can be a way for the public to pose questions on RMP.

- ❑ Establish a central phone number for display in the public service announcements that will allow you to field questions and respond or defer them to the proper locations (EPA, state, or stationary source representatives).

EVALUATING RMPlan DATA

- ❑ Develop skills in accessing and searching the database EPA will make available to you containing all applicable RMPlans for stationary sources.
- ❑ Request assistance from agencies in evaluating the information found in the RMPlans.
- ❑ Request assistance from the specific stationary sources in interpreting RMPlans.

USING RMPlan DATA

- ❑ Use worst case scenarios to pinpoint of potential problem areas in your community.
- ❑ Use alternative release scenarios to build decision trees for determining when to call for public shelter-in-place or evacuation.
- ❑ Use alternative release scenarios for preplanning evacuation routes.
- ❑ Use alternative release scenarios for planning drill scenarios and training exercises. Proactively approach stationary sources to perform these drills.
- ❑ Use five-year accident histories for planning realistic drill scenarios
- ❑ Use five-year accident histories for developing LEPC tabletop studies to help understand the best practices and weaknesses from past performance.
- ❑ Systematically begin requesting emergency response plans for review and arrange meetings to discuss them with stationary sources.
- ❑ Target those facilities whose RMPlans indicate more attention and request further information such as detailed alternative release scenarios and prevention program data.

- ❑ Approach facilities for assistance in planning out drill scenarios using actual data from their RMPlans.
- ❑ Consider revising the community emergency response plan based upon the reviews of the plans from each stationary source in your response area. Become a leader to help all the plans mesh seamlessly.

MAINTENANCE ACTIVITIES

- ❑ Maintain contact with the responsible RMP resource at stationary sources to stay aware of changes to their RMPlans and potential support they may require from your LEPC.
- ❑ Schedule drills with stationary sources and response agencies on a regular basis to continuously improve response methods and preplanning for realistic scenarios.
- ❑ Regularly repeat the shelter-in-place PSAs to maintain public awareness of their role in risk management.

7

Summary and Status of Revision to Regulations

This regulation has created a great deal of interest from both public action groups and industry. At this writing, several amendments are currently being addressed by the EPA. A summary of proposed amendments follows:

Use of a new industrial classification system is being considered:

- The recently adopted North American Industrial Classification System is proposed to replace the existing Standard Industrial Classification System.

Additional mandatory data elements are being considered:

- Method and description for latitude and longitude
- The Title V permit number
- The percentage weight of regulated toxic substances in mixtures reported in both the offsite consequence analyses and accident history
- NAICS code for any process that had a release reported in the five-year accident history section.

Optional data elements are being considered:

- LEPC identification information
- Stationary source e-mail address
- Stationary source homepage address
- Phone number for public inquiries
- Voluntary Protection Program status

Language changes are being considered:

- Prevention Program reporting language has been considered confusing.
- Changes have been offered to help companies define what a "process" is for the prevention program reporting requirements.

Confidential business information is being considered:

- Some wording regarding what can and cannot be considered confidential business information (CBI) is being drafted.
- This would have companies submit two RMPlans, one "sanitized" and one complete.
- The sanitized version would be made available to the public.
- LEPCs and the public would still be able to get a complete version with a written request.

General discussion of security due to terrorism:

- Recommendations have been made to EPA on the potential for increased terrorist activity targeted at chemical plants if all the data elements are made public.
- This topic is still under discussion at this time.
- Various techniques of controlling access to the data have been proposed.

Appendix A

Text of the EPA Risk Management Program Rule 40 CFR Part 68

68.1 Scope

This Part sets forth the list of regulated substances and thresholds, the petition process for adding or deleting substances to the list of regulated substances, the requirements for owners or operators of stationary sources concerning the prevention of accidental releases, and the State accidental release prevention programs approved under section 112(r). The list of substances, threshold quantities, and accident prevention regulations promulgated under this part do not limit in any way the general duty provisions under section 112(r)(1).

68.2 Stayed Provisions

(a) Notwithstanding any other provision of this part, the effectiveness of the following provisions is stayed from March 2, 1994 to December 22, 1997.

(a)(1) In Sec. 68.3, the definition of "stationary source," to the extent that such definition includes naturally occurring hydrocarbon reservoirs or transportation subject to oversight or regulation under a state natural gas or hazardous liquid program for which the state has in effect a certification to DOT under 49 U.S.C. 60105;

(a)(2) Section 68.115(b)(2) of this part, to the extent that such provision requires an owner or operator to treat as a regulated flammable substance:

(a)(2)(i) Gasoline, when in distribution or related storage for use as fuel for internal combustion engines;

(a)(2)(ii) Naturally occurring hydrocarbon mixtures prior to entry into a petroleum refining process unit or a natural gas processing plant. Naturally occurring hydrocarbon mixtures include any of the following: condensate, crude oil, field gas, and produced water, each as defined in paragraph (b) of this section;

(a)(2)(iii) Other mixtures that contain a regulated flammable substance and that do not have a National Fire Protection Association flammability hazard rating of 4, the definition of which is in the NFPA 704, Standard System for the Identification of the Fire Hazards of Materials, National Fire Protection Association, Quincy, MA, 1990, available from the National Fire Protection Association, 1 Batterymarch Park, Quincy, MA 02269-9101; and (a)(3) Section 68.130(a).

(b) From March 2, 1994 to December 22, 1997, the following definitions shall apply to the stayed provisions described in paragraph (a) of this section:

Condensate means hydrocarbon liquid separated from natural gas that condenses because of changes in temperature, pressure, or both, and remains liquid at standard conditions.

Crude oil means any naturally occurring, unrefined petroleum liquid.

Field gas means gas extracted from a production well before the gas enters a natural gas processing plant.

Natural gas processing plant means any processing site engaged in the extraction of natural gas liquids from field gas, fractionation of natural gas liquids to natural gas products, or both. A separator, dehydration unit, heater treater, sweetening unit, compressor, or similar equipment shall not be considered a "processing site" unless such equipment is physically located within a natural gas processing plant (gas plant) site. petroleum refining process unit means a process unit used in an establishment primarily engaged in petroleum refining as defined in the Standard Industrial Classification code for petroleum refining (2911) and used for the following: Producing transportation fuels (such as gasoline, diesel fuels, and jet fuels), heating fuels (such as kerosene, fuel gas distillate, and fuel oils), or lubricants; separating petroleum; or separating, cracking, reacting, or reforming intermediate petroleum streams. Examples of such units include, but are not limited to, petroleum based solvent units, alkylation units, catalytic hydrotreating, catalytic hydrorefining, catalytic hydrocracking, catalytic reforming, catalytic cracking, crude distillation, lube oil processing, hydrogen production, isomerization, polymerization, thermal processes, and blending, sweetening, and treating processes. Petroleum refining process units include sulfur plants. produced water means water extracted from the earth from an oil or natural gas production well, or that is separated from oil or natural gas after extraction.

68.3 Definitions.

For the purposes of this Part:

Accidental release means an unanticipated emission of a regulated substance or other extremely hazardous substance into the ambient air from a stationary source.

Act means the Clean Air Act as amended (42 U.S.C. 7401 et seq.)

Administrative controls mean written procedural mechanisms used for hazard control.

Administrator means the administrator of the U.S. Environmental protection Agency.

AIChE/CCPS means the American Institute of Chemical Engineers/Center for chemical Process Safety.

API means the American Petroleum Institute.

Article means a manufactured item, as defined under 29 CFR 1910.1200(b), that is formed to a specific shape or design during manufacture, that has end use functions dependent in whole or in part upon the shape or design during end use, and that does not release or otherwise result in exposure to a regulated substance under normal conditions of processing and use.

ASME means the American Society of Mechanical Engineers.

CAS means the Chemical Abstracts Service.

Catastrophic release means a major uncontrolled emission, fire, or explosion, involving one or more regulated substances that presents imminent and substantial endangerment to public health and the environment.

Classified information means "classified information" as defined in the Classified Information Procedures Act, 18 U.S.C. App. 3, section 1(a) as "any information or material that has been determined by the United States Government pursuant to an executive order, statute, or regulation, to require protection against unauthorized disclosure for reasons of national security."

Condensate means hydrocarbon liquid separated from natural gas that condenses due to changes in temperature, pressure, or both, and remains liquid at standard conditions.

Covered process means a process that has a regulated substance present in more than a threshold quantity as determined under §68.115.

Crude oil means any naturally occurring, unrefined petroleum liquid. Designated agency means the state, local, or Federal agency designated by the state under the provisions of §68.215(d).

DOT means the United States Department of Transportation.

Environmental receptor means natural areas such as national or state parks, forests, or monuments; officially designated wildlife sanctuaries, preserves, refuges, or areas; and Federal wilderness areas, that could be exposed at any time to toxic concentrations, radiant heat, or overpressure greater than or equal to the endpoints provided in §68.22(a), as a result of an accidental release and that can be identified on local U. S. Geological Survey maps.

Field gas means gas extracted from a production well before the gas enters a natural gas processing plant.

Hot work means work involving electric or gas welding, cutting, brazing, or similar flame or spark-producing operations.

Implementing agency means the state or local agency that obtains delegation for an accidental release prevention program under subpart E, 40 CFR part 63. The implementing agency may, but is not required to, be the state or local air permitting agency. If no state or local agency is granted delegation, EPA will be the implementing agency for that state.

Injury means any effect on a human that results either from direct exposure to toxic concentrations; radiant heat; or overpressures from accidental releases or from the direct consequences of a vapor cloud explosion (such as flying glass, debris, and other projectiles) from an accidental release and that requires medical treatment or hospitalization.

Major change means introduction of a new process, process equipment, or regulated substance, an alteration of process chemistry that results in any change to safe operating limits, or other alteration that introduces a new hazard.

Mechanical integrity means the process of ensuring that process equipment is fabricated from the proper materials of construction and is properly installed, maintained, and replaced to prevent failures and accidental releases.

Medical treatment means treatment, other than first aid, administered by a physician or registered professional personnel under standing orders from a physician.

Mitigation or mitigation system means specific activities, technologies, or equipment designed or deployed to capture or control substances upon loss of containment to minimize exposure of the public or the environment.

Passive mitigation means equipment, devices, or technologies that function without human, mechanical, or other energy input.

Active mitigation means equipment, devices, or technologies that need human, mechanical, or other energy input to function.

NFPA means the National Fire Protection Association.

Natural gas processing plant (gas plant) means any processing site engaged in the extraction of natural gas liquids from field gas, fractionation of mixed natural gas liquids to natural gas products, or both, classified as North American Industrial Classification System (NAICS) code 211112 (previously Standard Industrial Classification (SIC) code 1321).

Offsite means areas beyond the property boundary of the stationary source, and areas within the property boundary to which the public has routine and unrestricted access during or outside business hours.

OSHA means the U.S. Occupational Safety and Health Administration.

Owner or operator means any person who owns, leases, operates, controls, or supervises a stationary source.

Petroleum refining process unit means a process unit used in an establishment primarily engaged in petroleum refining as defined in NAICS code 32411 for petroleum refining (formerly SIC code 2911) and used for the following: Producing transportation fuels (such as gasoline, diesel fuels, and jet fuels), heating fuels (such as kerosene, fuel gas distillate, and fuel oils), or lubricants; Separating petroleum; or Separating, cracking, reacting, or reforming intermediate petroleum streams. Examples of such units include, but are not limited to, petroleum based solvent units, alkylation units, catalytic hydrotreating, catalytic hydrorefining, catalytic hydrocracking, catalytic reforming, catalytic cracking, crude distillation, lube oil processing, hydrogen production, isomerization, polymerization, thermal processes, and blending, sweetening, and treating processes. Petroleum refining process units include sulfur plants.

Population means the public.

Process means any activity involving a regulated substance including any use, storage, manufacturing, handling, or on-site movement of such substances, or combination of these activities. For the purposes of this definition, any group of vessels that are interconnected, or separate vessels that are located such that a regulated substance could be involved in a potential release, shall be considered a single process.

Produced water means water extracted from the earth from an oil or natural gas production well, or that is separated from oil or natural gas after extraction.

Public means any person except employees or contractors at the stationary source.

Public receptor means offsite residences, institutions (for example, schools, hospitals), industrial, commercial, and office buildings, parks, or recreational areas inhabited or occupied by the public at any time without restriction by the stationary source where members of the public could be exposed to toxic concentrations, radiant heat, or overpressure, as a result of an accidental release.

Regulated substance is any substance listed pursuant to section 112(r)(3) of the Clean Air Act as amended, in §68.130.

Replacement in kind means a replacement that satisfies the design specifications.

RMP means the risk management plan required under subpart G of this part.

SIC means Standard Industrial Classification.

Stationary source means any buildings, structures, equipment, installations, or substance emitting stationary activities which belong to the same industrial group, which are located on one or more contiguous properties, which are under the control of the same person (or persons under common control), and from which an accidental release may occur. The term "stationary source" does not apply to transportation, including storage incident to transportation, of any regulated substance or any other extremely hazardous substance under the provisions of this part. A stationary source includes transportation containers used for storage not incident to transportation and transportation containers connected to equipment at a stationary source for loading or unloading. Transportation includes, but is not limited to, transportation subject to oversight or regulation under 49 CFR parts 192, 193, or 195, or a state natural gas or hazardous liquid program for which the state has in effect a certification to DOT under 49 U.S.C. section 60105. A stationary source does not include naturally occurring hydrocarbon reservoirs. Properties shall not be considered contiguous solely because of a railroad or pipeline right-of-way.

Threshold quantity means the quantity specified for regulated substances pursuant to section 112(r)(5) of the Clean Air Act as amended, listed in §68.130 and determined to be present at a stationary source as specified in §68.115 of this Part.

Typical meteorological conditions means the temperature, wind speed, cloud cover, and atmospheric stability class, prevailing at the site based on data gathered at or near the site or from a local meteorological station.

Vessel means any reactor, tank, drum, barrel, cylinder, vat, kettle, boiler, pipe, hose, or other container.

Worst-case release means the release of the largest quantity of a regulated substance from a vessel or process line failure that results in the greatest distance to an endpoint defined in §68.22(a).

68.10 Applicability

(a) An owner or operator of a stationary source that has more than a threshold quantity of a regulated substance in a process, as determined under 68.115, shall comply with the requirements of this part no later than the latest of the following dates:

(a)(1) June 21, 1999;

(a)(2) Three years after the date on which a regulated substance is first listed under §68.130; or

(a)(3) The date on which a regulated substance is first present above a threshold quantity in a process.

(b) Program 1 eligibility requirements. A covered process is eligible for program 1 requirements as provided in §68.12(b) if it meets all of the following requirements:

(b)(1) For the five years prior to the submission of an RMP, the process has not had an accidental release of a regulated substance where exposure to the substance, its reaction products, overpressure generated by an explosion involving the substance, or radiant heat generated by a fire involving the substance led to any of the following offsite:

(b)(1)(i) Death;

(b)(1)(ii) Injury; or

(b)(1)(iii) Response or restoration activities for an exposure of an environmental receptor;

(b)(2) The distance to a toxic or flammable endpoint for a worst-case release assessment conducted under Subpart B and §68.25 is less than the distance to any public receptor, as defined in §68.30; and

(b)(3) Emergency response procedures have been coordinated between the stationary source and local emergency planning and response organizations.

(c) Program 2 eligibility requirements. A covered process is subject to program 2 requirements if it does not meet the eligibility requirements of either paragraph (b) or paragraph (d) of this section.

(d) Program 3 eligibility requirements. A covered process is subject to program 3 if the process does not meet the requirements of paragraph (b) of this section, and if either of the following conditions is met:

(d)(1) The process is in SIC code 2611, 2812, 2819, 2821, 2865, 2869, 2873, 2879, or 2911; or

(d)(2) The process is subject to the OSHA process safety management standard, 29 CFR 1910.119.

(e) If at any time a covered process no longer meets the eligibility criteria of its Program level, the owner or operator shall comply with the requirements of the new Program level that applies to the process and update the RMP as provided in §68.190.

(f) The provisions of this part shall not apply to an Outer Continental Shelf ("OCS") source, as defined in 40 CFR 55.2.

68.12 General Requirements

(a) General requirements. The owner or operator of a stationary source subject to this part shall submit a single RMP, as provided in §§68.150 to 68.185. The RMP shall include a registration that reflects all covered processes.

(b) Program 1 requirements. In addition to meeting the requirements of paragraph (a) of this section, the owner or operator of a stationary source with a process eligible for Program 1, as provided in §68.10(b), shall:

(b)(1) Analyze the worst-case release scenario for the process(es), as provided in §68.25; document that the nearest public receptor is beyond the distance to a toxic or flammable endpoint defined in §68.22(a); and submit in the RMP the worst-case release scenario as provided in §68.165;

(b)(2) Complete the five-year accident history for the process as provided in §68.42 of this part and submit it in the RMP as provided in §68.168;

(b)(3) Ensure that response actions have been coordinated with local emergency planning and response agencies; and

(b)(4) Certify in the RMP the following: "Based on the criteria in 40 CFR 68.10, the distance to the specified endpoint for the worst-case accidental release scenario for the following process(es) is less than the distance to the nearest public receptor: [list process(es)]. Within the past five years, the process(es) has (have) had no accidental release that caused offsite impacts provided in the risk management program rule (40 CFR 68.10(b)(1)). No additional measures are necessary to prevent offsite impacts from accidental releases. In the event of fire, explosion, or a release of a regulated substance from the process(es), entry within the distance to the specified endpoints may pose a danger to public emergency responders. Therefore, public emergency responders should not enter this area except as arranged with the emergency contact indicated in the RMP. The undersigned certifies that, to the best of my knowledge, information, and belief, formed after reasonable inquiry, the information submitted is true, accurate, and complete. [Signature, title, date signed]."

(c) Program 2 requirements. In addition to meeting the requirements of paragraph (a) of this section, the owner or operator of a stationary source with a process subject to Program 2, as provided in §68.10(c), shall:

(c)(1) Develop and implement a management system as provided in §68.15;

(c)(2) Conduct a hazard assessment as provided in §§68.20 through 68.42;

(c)(3) Implement the Program 2 prevention steps provided in §§68.48 through 68.60 or implement the Program 3 prevention steps provided in §§68.65 through 68.87;

(c)(4) Develop and implement an emergency response program as provided in §§68.90 to 68.95; and

(c)(5) Submit as part of the RMP the data on prevention program elements for Program 2 processes as provided in §68.170.

(d) Program 3 requirements. In addition to meeting the requirements of paragraph (a) of this section, the owner or operator of a stationary source with a process subject to Program 3, as provided in §68.10(d) shall:

(d)(1) Develop and implement a management system as provided in §68.15;

(d)(2) Conduct a hazard assessment as provided in §§68.20 through 68.42;

(d)(3) Implement the prevention requirements of §§68.65 through 68.87;

(d)(4) Develop and implement an emergency response program as provided in §§68.90 to 68.95 of this part; and

(d)(5) Submit as part of the RMP the data on prevention program elements for Program 3 processes as provided in §68.175.

68.15 Management

(a) The owner or operator of a stationary source with processes subject to Program 2 or Program 3 shall develop a management system to oversee the implementation of the risk management program elements.

(b) The owner or operator shall assign a qualified person or position that has the overall responsibility for the development, implementation, and integration of the risk management program elements.

(c) When responsibility for implementing individual requirements of this part is assigned to persons other than the person identified under paragraph (b) of this section, the names or positions of these people shall be documented and the lines of authority defined through an organization chart or similar document.

SUBPART B—HAZARD ASSESSMENT

68.20 Applicability

The owner or operator of a stationary source subject to this part shall prepare a worst-case release scenario analysis as provided in §68.25 of this part and complete the five-year accident history as provided in §68.42. The owner or operator of a Program 2 and 3 process must comply with all sections in this subpart for these processes.

68.22 Offsite Consequence Analysis Parameters

(a) *Endpoints.* For analyses of offsite consequences, the following endpoints shall be used:

(a)(1) *Toxics.* The toxic endpoints provided in Appendix A of this part.

(a)(2) *Flammables.* The endpoints for flammables vary according to the scenarios studied:

(a)(2)(i) *Explosion.* An overpressure of 1 psi.

(a)(2)(ii) *Radiant heat/exposure time.* A radiant heat of 5 kw/m^2 for 40 seconds.

(a)(2)(iii) *Lower flammability limit.* A lower flammability limit as provided in NFPA documents or other generally recognized sources.

(b) *Wind speed/atmospheric stability class.* For the worst-case release analysis, the owner or operator shall use a wind speed of 1.5 meters per second and F atmospheric stability class. If the owner or operator can demonstrate that local meteorological data applicable to the stationary source show a higher minimum wind speed or less stable atmosphere at all times during the previous three years, these minimums may be used. For analysis of alternative scenarios, the owner or operator may use the typical meteorological conditions for the stationary source.

(c) *Ambient temperature/humidity.* For worst-case release analysis of a regulated toxic substance, the owner or operator shall use the highest daily maximum temperature in the previous three years and average humidity for the site, based on temperature/humidity data gathered at the stationary source or at a local meteorological station; an owner or operator using the RMP Offsite Consequence Analysis Guidance may use 25°C and 50 percent humidity as values for these variables. For analysis of alternative scenarios, the owner or operator may use typical temperature/humidity data gathered at the stationary source or at a local meteorological station.

(d) *Height of release.* The worst-case release of a regulated toxic substance shall be analyzed assuming a ground level (0 feet) release. For an alternative scenario analysis of a regulated toxic substance, release height may be determined by the release scenario.

(e) *Surface roughness.* The owner or operator shall use either urban or rural topography, as appropriate. Urban means that there are many obstacles in the immediate area; obstacles include buildings or trees. Rural means there are no buildings in the immediate area and the terrain is generally flat and unobstructed.

(f) *Dense or neutrally buoyant gases.* The owner or operator shall ensure that tables or models used for dispersion analysis of regulated toxic substances appropriately account for gas density.

(g) *Temperature of released substance.* For worst case, liquids other than gases liquefied by refrigeration only shall be considered to be released at the highest daily maximum temperature, based on data for the previous three years appropriate for the stationary source, or at process temperature, whichever is higher. For alternative scenarios, substances may be considered to be released at a process or ambient temperature that is appropriate for the scenario.

68.25 Worst-Case Release Scenario Analysis

(a)(1) For Program 1 processes, one worst-case release scenario for each Program 1 process;

(a)(2) For Program 2 and 3 processes:

(a)(2)(i) One worst-case release scenario that is estimated to create the greatest distance in any direction to an endpoint provided in Appendix A of this part resulting from an accidental release of regulated toxic substances from covered processes under worst-case conditions defined in §68.22;

(a)(2)(ii) One worst-case release scenario that is estimated to create the greatest distance in any direction to an endpoint defined in §68.22(a) resulting from an accidental release of regulated flammable substances from covered processes under worst-case conditions defined in §68.22; and

(a)(2)(iii) Additional worst-case release scenarios for a hazard class if a worst-case release from another covered process at the stationary source potentially affects public receptors different from those potentially affected by the worst-case release scenario developed under paragraphs (a)(2)(i) or (a)(2)(ii) of this section.

(b) *Determination of worst-case release quantity.* The worst-case release quantity shall be the greater of the following:

(b)(1) For substances in a vessel, the greatest amount held in a single vessel, taking into account administrative controls that limit the maximum quantity; or

(b)(2) For substances in pipes, the greatest amount in a pipe, taking into account administrative controls that limit the maximum quantity.

(c) *Worst-case release scenario-toxic gases.*

(c)(1) For regulated toxic substances that are normally gases at ambient temperature and handled as a gas or as a liquid under pressure, the owner or operator shall assume that the quantity in the vessel or pipe, as determined under paragraph (b) of this section, is released as a gas over 10 minutes. The release rate shall be assumed to be the total quantity divided by 10 unless passive mitigation systems are in place.

(c)(2) For gases handled as refrigerated liquids at ambient pressure:

(c)(2)(i) If the released substance is not contained by passive mitigation systems or if the contained pool would have a depth of 1 cm or less, the owner or operator shall assume that the substance is released as a gas in 10 minutes;

(c)(2)(ii) If the released substance is contained by passive mitigation systems in a pool with a depth greater than 1 cm, the owner or operator may assume that the quantity in the vessel or pipe, as determined under paragraph (b) of this section, is spilled instantaneously to form a liquid pool. The volatilization rate (release rate) shall be calculated at the boiling point of the substance and at the conditions specified in paragraph (d) of this section.

(d) *Worst-case release scenario-toxic liquids.*

(d)(1) For regulated toxic substances that are normally liquids at ambient temperature, the owner or operator shall assume that the quantity in the vessel or pipe, as determined under paragraph (b) of this section, is spilled instantaneously to form a liquid pool.

(d)(1)(i) The surface area of the pool shall be determined by assuming that the liquid spreads to 1 centimeter deep unless passive mitigation systems are in place that serve to contain the spill and limit the surface area. Where passive mitigation is in place, the surface area of the contained liquid shall be used to calculate the volatilization rate.

(d)(1)(ii) If the release would occur onto a surface that is not paved or smooth, the owner or operator may take into account the actual surface characteristics.

(d)(2) The volatilization rate shall account for the highest daily maximum temperature occurring in the past three years, the temperature of the substance in the vessel, and the concentration of the substance if the liquid spilled is a mixture or solution.

(d)(3) The rate of release to air shall be determined from the volatilization rate of the liquid pool. The owner or operator may use the methodology in the RMP Offsite Consequence Analysis Guidance or any other publicly available techniques that account for the modeling conditions and are recognized by industry as applicable as part of current practices. proprietary models that account for the modeling conditions may be used provided the owner or operator allows the implementing agency access to the model and describes model features and differences from publicly available models to local emergency planners upon request.

(e) *Worst-case release scenario-flammables.* The owner or operator shall assume that the quantity of the substance, as determined under paragraph (b) of this section, vaporizes resulting in a vapor cloud explosion. A yield factor of 10 percent of the available energy released in the explosion shall be used to determine the distance to the explosion endpoint if the model used is based on TNT-equivalent methods.

(f) *Parameters to be applied.* The owner or operator shall use the parameters defined in §68.22 to determine distance to the endpoints. The owner or operator may use the methodology provided in the RMP Offsite Consequence Analysis Guidance or any commercially or publicly available air dispersion modeling techniques, provided the techniques account for the modeling conditions and are recognized by industry as applicable as part of current practices. Proprietary models that account for the modeling conditions may be used provided the owner or operator allows the implementing agency access to the model and describes model features and differences from publicly available models to local emergency planners upon request.

(g) *Consideration of passive mitigation.* Passive mitigation systems may be considered for the analysis of worst case provided that the mitigation system is capable of withstanding the release event triggering the scenario and would still function as intended.

(h) *Factors in selecting a worst-case scenario.* Notwithstanding the provisions of paragraph (b) of this section, the owner or operator shall select as the worst case for flammable regulated substances or the worst case for regulated toxic substances, a scenario based on the following factors if such a scenario would result in a greater distance to an endpoint defined in §68.22(a) beyond the stationary source boundary than the scenario provided under paragraph (b) of this section:

(h)(1) Smaller quantities handled at higher process temperature or pressure; and

(h)(2) Proximity to the boundary of the stationary source.

68.28 Alternative Release Scenario Analysis

(a) *The number of scenarios.* The owner or operator shall identify and analyze at least one alternative release scenario for each regulated toxic substance held in a covered process(es) and at least one alternative release scenario to represent all flammable substances held in covered processes.

(b) *Scenarios to consider.*

(b)(1) For each scenario required under paragraph (a) of this section, the owner or operator shall select a scenario:

(b)(1)(i) That is more likely to occur than the worst-case release scenario under §68.25; and

(b)(1)(ii) That will reach an endpoint offsite, unless no such scenario exists.

(b)(2) Release scenarios considered should include, but are not limited to, the following, where applicable:

(b)(2)(i) Transfer hose releases due to splits or sudden hose uncoupling;

(b)(2)(ii) Process piping releases from failures at flanges, joints, welds, valves and valve seals, and drains or bleeds;

(b)(2)(iii) Process vessel or pump releases due to cracks, seal failure, or drain, bleed, or plug failure;

(b)(2)(iv) Vessel overfilling and spill, or overpressurization and venting through relief valves or rupture disks; and

(b)(2)(v) Shipping container mishandling and breakage or puncturing leading to a spill.

(c) *Parameters to be applied.* The owner or operator shall use the appropriate parameters defined in §68.22 to determine distance to the endpoints. The owner or operator may use either the methodology provided in the RMP Offsite Consequence Analysis Guidance or any commercially or publicly available air dispersion modeling techniques, provided the techniques account for the specified modeling conditions and are recognized by industry as applicable as part of current practices. proprietary models that account for the modeling conditions may be used provided the owner or operator allows the implementing agency access to the model and describes model features and differences from publicly available models to local emergency planners upon request.

(d) *Consideration of mitigation.* Active and passive mitigation systems may be considered provided they are capable of withstanding the event that triggered the release and would still be functional.

(e) *Factors in selecting scenarios.* The owner or operator shall consider the following in selecting alternative release scenarios:

(e)(1) The five-year accident history provided in §68.42; and

(e)(2) Failure scenarios identified under §§68.50 or 68.67.

68.30 Defining Offsite Impacts—Population

(a) The owner or operator shall estimate in the RMP the population within a circle with its center at the point of the release and a radius determined by the distance to the endpoint defined in §68.22(a).

(b) *Population to be defined.* Population shall include residential population. The presence of institutions (schools, hospitals, prisons), parks and recreational areas, and major commercial, office, and industrial buildings shall be noted in the RMP.

(c) *Data sources acceptable.* The owner or operator may use the most recent Census data, or other updated information, to estimate the population potentially affected.

(d) Level of accuracy. Population shall be estimated to two significant digits.

68.33 Defining Offsite Impacts—Environment

(a) The owner or operator shall list in the RMP environmental receptors within a circle with its center at the point of the release and a radius determined by the distance to the endpoint defined in §68.22(a) of this part.

(b) *Data sources acceptable.* The owner or operator may rely on information provided on local U.S. Geological Survey maps or on any data source containing U.S.G.S. data to identify environmental receptors.

68.36 Review and Update

(a) The owner or operator shall review and update the offsite consequence analyses at least once every five years.

(b) If changes in processes, quantities stored or handled, or any other aspect of the stationary source might reasonably be expected to increase or decrease the distance to the endpoint by a factor of two or more, the owner or operator shall complete a revised analysis within six months of the change and submit a revised risk management plan as provided in §68.190.

68.39 Documentation

The owner or operator shall maintain the following records on the offsite consequence analyses:

(a) For worst-case scenarios, a description of the vessel or pipeline and substance selected as worst case, assumptions and parameters used, and the rationale for selection; assumptions shall include use of any administrative controls and any passive mitigation that were assumed to limit the quantity that could be released. Documentation shall include the anticipated effect of the controls and mitigation on the release quantity and rate.

(b) For alternative release scenarios, a description of the scenarios identified, assumptions and parameters used, and the rationale for the selection of specific scenarios; assumptions shall include use of any administrative controls and any mitigation that were assumed to limit the quantity that could be released. Documentation shall include the effect of the controls and mitigation on the release quantity and rate.

(c) Documentation of estimated quantity released, release rate, and duration of release.

(d) Methodology used to determine distance to endpoints.

(e) Data used to estimate population and environmental receptors potentially affected.

68.42 Five-Year Accident History

(a) The owner or operator shall include in the five-year accident history all accidental releases from covered processes that resulted in deaths, injuries, or significant property damage on site, or known offsite deaths, injuries, evacuations, sheltering in place, property damage, or environmental damage.

(b) *Data required.* For each accidental release included, the owner or operator shall report the following information:

(b)(1) Date, time, and approximate duration of the release;

(b)(2) Chemical(s) released;

(b)(3) Estimated quantity released in pounds;

(b)(4) The type of release event and its source;

(b)(5) Weather conditions, if known;

(b)(6) On-site impacts;

(b)(7) Known offsite impacts;

(b)(8) Initiating event and contributing factors if known;

(b)(9) Whether offsite responders were notified if known; and

(b)(10) Operational or process changes that resulted from investigation of the release.

(c) *Level of accuracy.* Numerical estimates may be provided to two significant digits.

SUBPART C—PROGRAM 2 PREVENTION PROGRAM

68.48 Safety Information

(a) The owner or operator shall compile and maintain the following up-to-date safety information related to the regulated substances, processes, and equipment:

(a)(1) Material Safety Data Sheets that meet the requirements of 29 CFR 1910.1200(g);

(a)(2) Maximum intended inventory of equipment in which the regulated substances are stored or processed;

(a)(3) Safe upper and lower temperatures, pressures, flows, and compositions;

(a)(4) Equipment specifications; and

(a)(5) Codes and standards used to design, build, and operate the process.

(b) The owner or operator shall ensure that the process is designed in compliance with recognized and generally accepted good engineering practices. Compliance with Federal or state regulations that address industry-specific safe design or with industry-specific design codes and standards may be used to demonstrate compliance with this paragraph.

(c) The owner or operator shall update the safety information if a major change occurs that makes the information inaccurate.

68.50 Hazard Review

(a) The owner or operator shall conduct a review of the hazards associated with the regulated substances, process, and procedures. The review shall identify the following:

(a)(1) The hazards associated with the process and regulated substances;

(a)(2) Opportunities for equipment malfunctions or human errors that could cause an accidental release;

(a)(3) The safeguards used or needed to control the hazards or prevent equipment malfunction or human error; and

(a)(4) Any steps used or needed to detect or monitor releases.

(b) The owner or operator may use checklists developed by persons or organizations knowledgeable about the process and equipment as a guide to conducting the review. For processes designed to meet industry standards or Federal or state design rules, the hazard review shall, by inspecting all equipment, determine whether the process is designed, fabricated, and operated in accordance with the applicable standards or rules.

(c) The owner or operator shall document the results of the review and ensure that problems identified are resolved in a timely manner.

(d) The review shall be updated at least once every five years. The owner or operator shall also conduct reviews whenever a major change in the process occurs; all issues identified in the review shall be resolved before startup of the changed process.

68.52 Operating Procedures

(a) The owner or operator shall prepare written operating procedures that provide clear instructions or steps for safely conducting activities associated with each covered process consistent with the safety information for that process. Operating procedures or instructions provided by equipment manufacturers or developed by persons or organizations knowledgeable about the process and equipment may be used as a basis for a stationary source's operating procedures.

(b) The procedures shall address the following:

(b)(1) Initial startup;

(b)(2) Normal operations;

(b)(3) Temporary operations;

(b)(4) Emergency shutdown and operations;

(b)(5) Normal shutdown;

(b)(6) Startup following a normal or emergency shutdown or a major change that requires a hazard review;

(b)(7) Consequences of deviations and steps required to correct or avoid deviations; and

(b)(8) Equipment inspections.

(c) The owner or operator shall ensure that the operating procedures are updated, if necessary, whenever a major change occurs and prior to startup of the changed process.

68.54 Training

(a) The owner or operator shall ensure that each employee presently operating a process, and each employee newly assigned to a covered process have been trained or tested competent in the operating procedures provided in §68.52 that pertain to their duties. For those employees already operating a process on June 21, 1999, the owner or operator may certify in writing that the employee has the required knowledge, skills, and abilities to safely carry out the duties and responsibilities as provided in the operating procedures.

(b) *Refresher training.* Refresher training shall be provided at least every three years, and more often if necessary, to each employee operating a process to ensure that the employee understands and adheres to the current operating procedures of the process. The owner or operator, in consultation with the employees operating the process, shall determine the appropriate frequency of refresher training.

(c) The owner or operator may use training conducted under Federal or state regulations or under industry-specific standards or codes or training conducted by covered process equipment vendors to demonstrate compliance with this section to the extent that the training meets the requirements of this section.

(d) The owner or operator shall ensure that operators are trained in any updated or new procedures prior to startup of a process after a major change.

68.56 Maintenance

(a) The owner or operator shall prepare and implement procedures to maintain the on-going mechanical integrity of the process equipment. The owner or operator may use procedures or instructions provided by covered process equipment vendors or procedures in Federal or state regulations or industry codes as the basis for stationary source maintenance procedures.

(b) The owner or operator shall train or cause to be trained each employee involved in maintaining the on-going mechanical integrity of the process. To ensure that the employee can perform the job tasks in a safe manner, each such employee shall be trained in the hazards of the process, in how to avoid or correct unsafe conditions, and in the procedures applicable to the employee's job tasks.

(c) Any maintenance contractor shall ensure that each contract maintenance employee is trained to perform the maintenance procedures developed under paragraph (a) of this section.

(d) The owner or operator shall perform or cause to be performed inspections and tests on process equipment. Inspection and testing procedures shall follow recognized and generally accepted good engineering practices. The frequency of inspections and tests of process equipment shall be consistent with applicable manufacturers' recommendations, industry standards or codes, good engineering practices, and prior operating experience.

68.58 Compliance Audits

(a) The owner or operator shall certify that they have evaluated compliance with the provisions of this subpart at least every three years to verify that the procedures and practices developed under the rule are adequate and are being followed.

(b) The compliance audit shall be conducted by at least one person knowledgeable in the process.

(c) The owner or operator shall develop a report of the audit findings.

(d) The owner or operator shall promptly determine and document an appropriate response to each of the findings of the compliance audit and document that deficiencies have been corrected.

(e) The owner or operator shall retain the two (2) most recent compliance audit reports. This requirement does not apply to any compliance audit report that is more than five years old.

68.60 Incident Investigation

(a) The owner or operator shall investigate each incident which resulted in, or could reasonably have resulted in a catastrophic release.

(b) An incident investigation shall be initiated as promptly as possible, but not later than 48 hours following the incident.

(c) A summary shall be prepared at the conclusion of the investigation which includes at a minimum:

(c)(1) Date of incident;

(c)(2) Date investigation began;

(c)(3) A description of the incident;

(c)(4) The factors that contributed to the incident; and,

(c)(5) Any recommendations resulting from the investigation.

(d) The owner or operator shall promptly address and resolve the investigation findings and recommendations. Resolutions and corrective actions shall be documented.

(e) The findings shall be reviewed with all affected personnel whose job tasks are affected by the findings.

(f) Investigation summaries shall be retained for five years.

SUBPART D—PROGRAM 3 PREVENTION PROGRAM

68.65 Process Safety Information

(a) In accordance with the schedule set forth in §68.67, the owner or operator shall complete a compilation of written process safety information before conducting any process hazard analysis required by the rule. The compilation of written process safety information is to enable the owner or operator and the employees involved in operating the process to identify and understand the hazards posed by those processes involving regulated substances. This process safety information shall include information pertaining to the hazards of the regulated substances used or produced by the process, information pertaining to the technology of the process, and information pertaining to the equipment in the process.

(b) Information pertaining to the hazards of the regulated substances in the process. This information shall consist of at least the following:

(b)(1) Toxicity information;

(b)(2) Permissible exposure limits;

(b)(3) Physical data;

(b)(4) Reactivity data:

(b)(5) Corrosivity data;

(b)(6) Thermal and chemical stability data; and

(b)(7) Hazardous effects of inadvertent mixing of different materials that could foreseeably occur. Note to paragraph (b): Material Safety Data Sheets meeting the requirements of 29 CFR 1910.1200(g) may be used to comply with this requirement to the extent they contain the information required by this subparagraph.

(c) Information pertaining to the technology of the process.

(c)(1) Information concerning the technology of the process shall include at least the following:

(c)(1)(i) A block flow diagram or simplified process flow diagram;

(c)(1)(ii) Process chemistry;

(c)(1)(iii) Maximum intended inventory;

(c)(1)(iv) Safe upper and lower limits for such items as

temperatures, pressures, flows or compositions; and,

(c)(1)(v) An evaluation of the consequences of deviations.

(c)(2) Where the original technical information no longer exists, such information may be developed in conjunction with the process hazard analysis in sufficient detail to support the analysis.

(d) Information pertaining to the equipment in the process.

(d)(1) Information pertaining to the equipment in the process shall include:

(d)(1)(i) Materials of construction;

(d)(1)(ii) Piping and instrument diagrams (P&ID's);

(d)(1)(iii) Electrical classification;

(d)(1)(iv) Relief system design and design basis;

(d)(1)(v) Ventilation system design;

(d)(1)(vi) Design codes and standards employed;

(d)(1)(vii) Material and energy balances for processes built after June 21, 1999; and

(d)(1)(viii) Safety systems (for example interlocks, detection or suppression systems).

(d)(2) The owner or operator shall document that equipment complies with recognized and generally accepted good engineering practices.

(d)(3) For existing equipment designed and constructed in accordance with codes, standards, or practices that are no longer in general use, the owner or operator shall determine and document that the equipment is designed, maintained, inspected, tested, and operating in a safe manner.

68.67 Process Hazard Analysis

(a) The owner or operator shall perform an initial process hazard analysis (hazard evaluation) on processes covered by this part. The process hazard analysis shall be appropriate to the complexity of the process and shall identify, evaluate, and control the hazards involved in the process. The owner or operator shall determine and document the priority order for conducting process hazard analyses based on a rationale which includes such considerations as extent of the process hazards, number of potentially affected employees, age of the process, and operating history of the process. The process hazard analysis shall be conducted as soon as possible, but not later than June 21, 1999. Process hazards analyses completed to comply with 29 CFR 1910.119(e) are acceptable as initial process hazards analyses. These process hazard analyses shall be updated and revalidated, based on their completion date.

(b) The owner or operator shall use one or more of the following methodologies that are appropriate to determine and evaluate the hazards of the process being analyzed.

(b)(1) What-If;

(b)(2) Checklist;

(b)(3) What-If/Checklist;

(b)(4) Hazard and Operability Study (HAZOP);

(b)(5) Failure Mode and Effects Analysis (FMEA);

(b)(6) Fault Tree Analysis; or

(b)(7) An appropriate equivalent methodology.

(c) The process hazard analysis shall address:

(c)(1) The hazards of the process;

(c)(2) The identification of any previous incident which had a likely potential for catastrophic consequences.

(c)(3) Engineering and administrative controls applicable to the hazards and their interrelationships such as appropriate application of detection methodologies to provide early warning of releases. (Acceptable detection methods might include process monitoring and control instrumentation with alarms, and detection hardware such as hydrocarbon sensors.);

(c)(4) Consequences of failure of engineering and administrative controls;

(c)(5) Stationary source siting;

(c)(6) Human factors; and

(c)(7) A qualitative evaluation of a range of the possible safety and health effects of failure of controls.

(d) The process hazard analysis shall be performed by a team with expertise in engineering and process operations, and the team shall include at least one employee who has experience and knowledge specific to the process being evaluated. Also, one member of the team must be knowledgeable in the specific process hazard analysis methodology being used.

(e) The owner or operator shall establish a system to promptly address the team's findings and recommendations; assure that the recommendations are resolved in a timely manner and that the resolution is documented; document what actions are to be taken; complete actions as soon as possible; develop a written schedule of when these actions are to be completed; communicate the actions to operating, maintenance and other employees whose work assignments are in the process and who may be affected by the recommendations or actions.

(f) At least every five (5) years after the completion of the initial process hazard analysis, the process hazard analysis shall be updated and revalidated by a team meeting the requirements in paragraph (d) of this section, to assure that the process hazard analysis is consistent with the current process. Updated and revalidated process hazard analyses completed to comply with 29 CFR 1910.119(e) are acceptable to meet the requirements of this paragraph.

(g) The owner or operator shall retain process hazards analyses and updates or revalidations for each process covered by this section, as well as the documented resolution of recommendations described in paragraph (e) of this section for the life of the process.

68.69 Operating Procedures

(a) The owner or operator shall develop and implement written operating procedures that provide clear instructions for safely conducting activities involved in each covered process consistent with the process safety information and shall address at least the following elements.

(a)(1) Steps for each operating phase:

(a)(1)(i) Initial startup;

(a)(1)(ii) Normal operations;

(a)(1)(iii) Temporary operations;

(a)(1)(iv) Emergency shutdown including the conditions under which emergency shutdown is required, and the assignment of shutdown responsibility to qualified operators to ensure that emergency shutdown is executed in a safe and timely manner.

(a)(1)(v) Emergency operations;

(a)(1)(vi) Normal shutdown; and,

(a)(1)(vii) Startup following a turnaround, or after an emergency shutdown.

(a)(2) Operating limits:

(a)(2)(i) Consequences of deviation; and

(a)(2)(ii) Steps required to correct or avoid deviation.

(a)(3) Safety and health considerations:

(a)(3)(i) Properties of, and hazards presented by, the chemicals used in the process;

(a)(3)(ii) Precautions necessary to prevent exposure, including engineering controls, administrative controls, and personal protective equipment;

(a)(3)(iii) Control measures to be taken if physical contact or airborne exposure occurs;

(a)(3)(iv) Quality control for raw materials and control of hazardous chemical inventory levels; and,

(a)(3)(v) Any special or unique hazards.

(a)(4) Safety systems and their functions.

(b) Operating procedures shall be readily accessible to employees who work in or maintain a process.

(c) The operating procedures shall be reviewed as often as necessary to assure that they reflect current operating practice, including changes that result from changes in process chemicals, technology, and equipment, and changes to stationary sources. The owner or operator shall certify annually that these operating procedures are current and accurate.

(d) The owner or operator shall develop and implement safe work practices to provide for the control of hazards during operations such as lockout/tagout; confined space entry; opening process equipment or piping; and control over entrance into a stationary source by maintenance, contractor, laboratory, or other support personnel. These safe work practices shall apply to employees and contractor employees.

68.71 Training

(a) *Initial training.*

(a)(1) Each employee presently involved in operating a process, and each employee before being involved in operating a newly assigned process, shall be trained in an overview of the process and in the operating procedures as specified in §68.69. The training shall include emphasis on the specific safety and health hazards, emergency operations including shutdown, and safe work practices applicable to the employee's job tasks.

(a)(2) In lieu of initial training for those employees already involved in operating a process on June 21, 1999 an owner or operator may certify in writing that the employee has the required knowledge, skills, and abilities to safely carry out the duties and responsibilities as specified in the operating procedures.

(b) *Refresher training.* Refresher training shall be provided at least every three years, and more often if necessary, to each employee involved in operating a process to assure that the employee understands and adheres to the current operating procedures of the process. The owner or operator, in consultation with the employees involved in operating the process, shall determine the appropriate frequency of refresher training.

(c) *Training documentation.* The owner or operator shall ascertain that each employee involved in operating a process has received and understood the training required by this paragraph. The owner or operator shall prepare a record which contains the identity of the employee, the date of training, and the means used to verify that the employee understood the training.

68.73 Mechanical Integrity

(a) *Application.* Paragraphs (b) through (f) of this section apply to the following process equipment:

(a)(1) Pressure vessels and storage tanks;

(a)(2) Piping systems (including piping components such as valves);

(a)(3) Relief and vent systems and devices;

(a)(4) Emergency shutdown systems;

(a)(5) Controls (including monitoring devices and sensors, alarms, and interlocks) and,

(a)(6) Pumps.

(b) *Written procedures.* The owner or operator shall establish and implement written procedures to maintain the on-going integrity of process equipment.

(c) *Training for process maintenance activities.* The owner or operator shall train each employee involved in maintaining the on-going integrity of process equipment in an overview of that process and its hazards and in the procedures applicable to the employee's job tasks to assure that the employee can perform the job tasks in a safe manner.

(d) *Inspection and testing.*

(d)(1) Inspections and tests shall be performed on process equipment.

(d)(2) Inspection and testing procedures shall follow recognized and generally accepted good engineering practices.

(d)(3) The frequency of inspections and tests of process equipment shall be consistent with applicable manufacturers' recommendations and good engineering practices, and more frequently if determined to be necessary by prior operating experience.

(d)(4) The owner or operator shall document each inspection and test that has been performed on process equipment. The documentation shall identify the date of the inspection or test, the name of the person who performed the inspection or test, the serial number or other identifier of the equipment on which the inspection or test was performed, a description of the inspection or test performed, and the results of the inspection or test.

(e) *Equipment deficiencies.* The owner or operator shall correct deficiencies in equipment that are outside acceptable limits (defined by the process safety information in §68.65) before further use or in a safe and timely manner when necessary means are taken to assure safe operation.

(f) *Quality assurance.*

(f)(1) In the construction of new plants and equipment, the owner or operator shall assure that equipment as it is fabricated is suitable for the process application for which they will be used.

(f)(2) Appropriate checks and inspections shall be performed to assure that equipment is installed properly and consistent with design specifications and the manufacturer's instructions.

(f)(3) The owner or operator shall assure that maintenance materials, spare parts and equipment are suitable for the process application for which they will be used.

68.75 Management of Change

(a) The owner or operator shall establish and implement written procedures to manage changes (except for "replacements in kind") to process chemicals, technology, equipment, and procedures; and, changes to stationary sources that affect a covered process.

(b) The procedures shall assure that the following considerations are addressed prior to any change:

(b)(1) The technical basis for the proposed change;

(b)(2) Impact of change on safety and health;

(b)(3) Modifications to operating procedures;

(b)(4) Necessary time period for the change; and,

(b)(5) Authorization requirements for the proposed change.

(c) Employees involved in operating a process and maintenance and contract employees whose job tasks will be affected by a change in the process shall be informed of, and trained in, the change prior to start-up of the process or affected part of the process.

(d) If a change covered by this paragraph results in a change in the process safety information required by §68.65 of this part, such information shall be updated accordingly.

(e) If a change covered by this paragraph results in a change in the operating procedures or practices required by §68.69, such procedures or practices shall be updated accordingly.

68.77 Pre-Startup Review

(a) The owner or operator shall perform a pre-startup safety review for new stationary sources and for modified stationary sources when the modification is significant enough to require a change in the process safety information.

(b) The pre-startup safety review shall confirm that prior to the introduction of regulated substances to a process:

(b)(1) Construction and equipment is in accordance with design specifications;

(b)(2) Safety, operating, maintenance, and emergency procedures are in place and are adequate;

(b)(3) For new stationary sources, a process hazard analysis has been performed and recommendations have been resolved or implemented before startup; and modified stationary sources meet the requirements contained in management of change, §68.75.

(b)(4) Training of each employee involved in operating a process has been completed.

68.79 Compliance Audits

(a) The owner or operator shall certify that they have evaluated compliance with the provisions of this section at least every three years to verify that the procedures and practices developed under the standard are adequate and are being followed.

(b) The compliance audit shall be conducted by at least one person knowledgeable in the process.

(c) A report of the findings of the audit shall be developed.

(d) The owner or operator shall promptly determine and document an appropriate response to each of the findings of the compliance audit, and document that deficiencies have been corrected.

(e) The owner or operator shall retain the two (2) most recent compliance audit reports.

68.81 Incident Investigation

(a) The owner or operator shall investigate each incident which resulted in, or could reasonably have resulted in a catastrophic release of a regulated substance.

(b) An incident investigation shall be initiated as promptly as possible, but not later than 48 hours following the incident.

(c) An incident investigation team shall be established and consist of at least one person knowledgeable in the process involved, including a contract employee if the incident involved work of the contractor, and other persons with appropriate knowledge and experience to thoroughly investigate and analyze the incident.

(d) A report shall be prepared at the conclusion of the investigation which includes at a minimum:

(d)(1) Date of incident;

(d)(2) Date investigation began;

(d)(3) A description of the incident;

(d)(4) The factors that contributed to the incident; and,

(d)(5) Any recommendations resulting from the investigation.

(e) The owner or operator shall establish a system to promptly address and resolve the incident report findings and recommendations. Resolutions and corrective actions shall be documented.

(f) The report shall be reviewed with all affected personnel whose job tasks are relevant to the incident findings including contract employees where applicable.

(g) Incident investigation reports shall be retained for five years.

68.83 Employee Participation

(a) The owner or operator shall develop a written plan of action regarding the implementation of the employee participation required by this section.

(b) The owner or operator shall consult with employees and their representatives on the conduct and development of process hazards analyses and on the development of the other elements of process safety management in this rule.

(c) The owner or operator shall provide to employees and their representatives access to process hazard analyses and to all other information required to be developed under this rule.

68.85 Hot Work Permit

(a) The owner or operator shall issue a hot work permit for hot work operations conducted on or near a covered process.

(b) The permit shall document that the fire prevention and protection requirements in 29 CFR 1910.252(a) have been implemented prior to beginning the hot work operations; it shall indicate the date(s) authorized for hot work; and identify the object on which hot work is to be performed. The permit shall be kept on file until completion of the hot work operations.

68.87 Contractors

(a) *Application.* This section applies to contractors performing maintenance or repair, turnaround, major renovation, or specialty work on or adjacent to a covered process. It does not apply to contractors providing incidental services which do not influence process safety, such as janitorial work, food and drink services, laundry, delivery or other supply services.

(b) Owner or operator responsibilities. (1) The owner or operator, when selecting a contractor, shall obtain and evaluate information regarding the contract owner or operator's safety performance and programs.

(b)(2) The owner or operator shall inform contract owner or operator of the known potential fire, explosion, or toxic release hazards related to the contractor's work and the process.

(b)(3) The owner or operator shall explain to the contract owner or operator the applicable provisions of subpart E of this part.

(b)(4) The owner or operator shall develop and implement safe work practices consistent with §68.69(d), to control the entrance, presence, and exit of the contract owner or operator and contract employees in covered process areas.

(b)(5) The owner or operator shall periodically evaluate the performance of the contract owner or operator in fulfilling their obligations as specified in paragraph (c) of this section.

(c) *Contract owner or operator responsibilities.*

(c)(1) The contract owner or operator shall assure that each contract employee is trained in the work practices necessary to safely perform his/her job.

(c)(2) The contract owner or operator shall assure that each contract employee is instructed in the known potential fire, explosion, or toxic release hazards related to his/her job and the process, and the applicable provisions of the emergency action plan.

(c)(3) The contract owner or operator shall document that each contract employee has received and understood the training required by this section. The contract owner or operator shall prepare a record which contains the identity of the contract employee, the date of training, and the means used to verify that the employee understood the training.

(c)(4) The contract owner or operator shall assure that each contract employee follows the safety rules of the stationary source including the safe work practices required by §68.69(d).

(c)(5) The contract owner or operator shall advise the owner or operator of any unique hazards presented by the contract owner or operator's work, or of any hazards found by the contract owner or operator's work.

SUBPART E—EMERGENCY RESPONSE

68.90 Applicability.

(a) Except as provided in paragraph (b) of this section, the owner or operator of a stationary source with Program 2 and Program 3 processes shall comply with the requirements of §68.95.

(b) The owner or operator of stationary source whose employees will not respond to accidental releases of regulated substances need not comply with §68.95 of this part provided that they meet the following:

(b)(1) For stationary sources with any regulated toxic substance held in a process above the threshold quantity, the stationary source is included in the community emergency response plan developed under 42 U.S.C. 11003;

(b)(2) For stationary sources with only regulated flammable substances held in a process above the threshold quantity, the owner or operator has coordinated response actions with the local fire department; and

(b)(3) Appropriate mechanisms are in place to notify emergency responders when there is a need for a response.

68.95 Emergency Response Program

(a) The owner or operator shall develop and implement an emergency response program for the purpose of protecting public health and the environment. Such program shall include the following elements:

(a)(1) An emergency response plan, which shall be maintained at the stationary source and contain at least the following elements:

(a)(1)(i) Procedures for informing the public and local emergency response agencies about accidental releases;

(a)(1)(ii) Documentation of proper first-aid and emergency medical treatment necessary to treat accidental human exposures; and

(a)(1)(iii) Procedures and measures for emergency response after an accidental release of a regulated substance;

(a)(2) Procedures for the use of emergency response equipment and for its inspection, testing, and maintenance;

(a)(3) Training for all employees in relevant procedures; and (a)(4) Procedures to review and update, as appropriate, the emergency response plan to reflect changes at the stationary source and ensure that employees are informed of changes.

(b) A written plan that complies with other Federal contingency plan regulations or is consistent with the approach in the National Response Team's Integrated Contingency Plan Guidance ("One Plan") and that, among other matters, includes the elements provided in paragraph (a) of this section, shall satisfy the requirements of this section if the owner or operator also complies with paragraph (c) of this section.

(c) The emergency response plan developed under paragraph (a)(1) of this section shall be coordinated with the community emergency response plan developed under 42 U.S.C. 11003. Upon request of the local emergency planning committee or emergency response officials, the owner or operator shall promptly provide to the local emergency response officials information necessary for developing and implementing the community emergency response plan.

SUBPART F—REGULATED SUBSTANCES FOR ACCIDENTAL RELEASE PREVENTION

68.100 Purpose

This subpart designates substances to be listed under section 112(r)(3), (4), and (5) of the Clean Air Act, as amended, identifies their threshold quantities, and establishes the requirements for petitioning to add or delete substances from the list.

68.115 Threshold Determination

(a) A threshold quantity of a regulated substance listed in §68.130 is present at a stationary source if the total quantity of the regulated substance contained in a process exceeds the threshold.

(b) For the purposes of determining whether more than a threshold quantity of a regulated substance is present at the stationary source, the following exemptions apply:

(b)(1) *Concentrations of a regulated toxic substance in a mixture.* If a regulated substance is present in a mixture and the concentration of the substance is below one percent by weight of the mixture, the amount of the substance in the mixture need not be considered when determining whether more than a threshold quantity is present at the stationary source. Except for oleum, toluene 2,4-diisocyanate, toluene 2,6-diisocyanate, and toluene diisocyanate (unspecified isomer), if the concentration of the regulated substance in the mixture is one percent or greater by weight, but the owner or operator can demonstrate that the partial pressure of the regulated substance in the mixture (solution) under handling or storage conditions in any portion of the process is less than 10 millimeters of mercury (mm Hg), the amount of the substance in the mixture in that portion of the process need not be considered when determining whether more than a threshold quantity is present at the stationary source. The owner or operator shall document this partial pressure measurement or estimate.

(b)(2) *Concentrations of a regulated flammable substance in a mixture.*

(B)(2)(i) *General provision.* If a regulated substance is present in a mixture and the concentration of the substance is below one percent by weight of the mixture, the mixture need not be considered when determining whether more than a threshold quantity of the regulated substance is present at the stationary source. Except as provided in paragraph (b)(2)(ii) and (iii) of this section, if the concentration of the substance is one percent or greater by weight of the mixture, then, for purposes of determining whether a threshold quantity is present at the stationary source, the entire weight of the mixture shall be treated as the regulated substance unless the owner or operator can demonstrate that the mixture itself does not have a National Fire Protection Association flammability hazard rating of 4. The demonstration shall be in accordance with the definition of flammability hazard rating 4 in the NFPA 704, Standard System for the Identification of the Hazards of Materials for Emergency Response, National Fire Protection Association, Quincy, MA, 1996. Available from the National Fire Protection Association, 1 Batterymarch Park, Quincy, MA 02269-9101. This incorporation by reference was approved by the Director of the Federal Register in accordance with 5 U.S.C. 552(a) and 1 CFR part 51. Copies may be inspected at the Environmental Protection Agency Air Docket (6102), Attn: Docket No. A-96-O8, Waterside Mall, 401 M. St. SW., Washington D.C.; or at the Office of Federal Register at 800 North Capitol St., NW, Suite 700, Washington, D.C. Boiling point and flash point shall be defined and determined in accordance with NFPA 30, Flammable and Combustible Liquids Code, National Fire Protection Association, Quincy, MA, 1996.

Available from the National Fire Protection Association, 1 Batterymarch Park, Quincy, MA 02269-9101. This incorporation by reference was approved by the Director of the Federal Register in accordance with 5 U.S.C. 552(a) and 1 CFR part 51. Copies may be inspected at the Environmental Protection Agency Air Docket (6102), Attn: Docket No. A-96-08, Waterside Mall, 401 M. St. SW., Washington D.C.; or at the Office of Federal Register at 800 North Capitol St., NW, Suite 700, Washington, D.C. The owner or operator shall document the National Fire protection Association flammability hazard rating.

(b)(2)(ii) *Gasoline.* Regulated substances in gasoline, when in distribution or related storage for use as fuel for internal combustion engines, need not be considered when determining whether more than a threshold quantity is present at a stationary source.

(b)(2)(iii) *Naturally occurring hydrocarbon mixtures.* Prior to entry into a natural gas processing plant or a petroleum refining process unit, regulated substances in naturally occurring hydrocarbon mixtures need not be considered when determining whether more than a threshold quantity is present at a stationary source. Naturally occurring hydrocarbon mixtures include any combination of the following: condensate, crude oil, field gas, and produced water, each as defined in Sec. 68.3 of this part.

(b)(3) Concentrations of a regulated explosive substance in a mixture. Mixtures of Division 1.1 explosives listed in 49 CFR 172.101 (Hazardous Materials Table) and other explosives need not be included when determining whether a threshold quantity is present in a process, when the mixture is intended to be used on- site in a non-accidental release in a manner consistent with applicable BATF regulations. Other mixtures of Division 1.1 explosives listed in 49 CFR 172.101 and other explosives shall be included in determining whether more than a threshold quantity is present in a process if such mixtures would be treated as Division 1.1 explosives under 49 CFR Parts 172 and 173.

(b)(4) *Articles.* Regulated substances contained in articles need not be considered when determining whether more than a threshold quantity is present at the stationary source.

(b)(5) *Uses.* Regulated substances, when in use for the following purposes, need not be included in determining whether more than a threshold quantity is present at the stationary source:

(b)(5)(i) Use as a structural component of the stationary source;

(b)(5)(ii) Use of products for routine janitorial maintenance;

(b)(5)(iii) Use by employees of foods, drugs, cosmetics, or other personal items containing the regulated substance; and

(b)(5)(iv) Use of regulated substances present in process water or non-contact cooling water as drawn from the environment or municipal sources, or use of regulated substances present in air used either as compressed air or as part of combustion.

(b)(6) *Activities in Laboratories.* If a regulated substance is manufactured, processed, or used in a laboratory at a stationary source under the supervision of a technically qualified individual as defined in §720.3(ee) of this chapter, the quantity of the substance need not be considered in determining whether a threshold quantity is present. This exemption does not apply to:

(b)(6)(i) Specialty chemical production;

(b)(6)(ii) Manufacture, processing, or use of substances in pilot plant scale operations; and

(b)(6)(iii) Activities conducted outside the laboratory.

68.120 Petition Process

(a) Any person may petition the Administrator to modify, by addition or deletion, the list of regulated substances identified in §68.130. Based on the information presented by the petitioner, the Administrator may grant or deny a petition.

(b) A substance may be added to the list if, in the case of an accidental release, it is known to cause or may be reasonably anticipated to cause death, injury, or serious adverse effects to human health or the environment.

(c) A substance may be deleted from the list if adequate data on the health and environmental effects of the substance are available to determine that the substance, in the case of an accidental release, is not known to cause and may not be reasonably anticipated to cause death, injury, or serious adverse effects to human health or the environment.

(d) No substance for which a national primary ambient air quality standard has been established shall be added to the list. No substance regulated under Title VI of the Clean Air Act, as amended, shall be added to the list.

(e) The burden of proof is on the petitioner to demonstrate that the criteria for addition and deletion are met. A petition will be denied if this demonstration is not made.

(f) The Administrator will not accept additional petitions on the same substance following publication of a final notice of the decision to grant or deny a petition, unless new data becomes available that could significantly affect the basis for the decision.

(g) Petitions to modify the list of regulated substances must contain the following:

(g)(1) Name and address of the petitioner and a brief description of the organization(s) that the petitioner represents, if applicable;

(g)(2) Name, address, and telephone number of a contact person for the petition;

(g)(3) Common chemical name(s), common synonym(s), Chemical Abstracts Service number, and chemical formula and structure;

(g)(4) Action requested (add or delete a substance);

(g)(5) Rationale supporting the petitioner's position; that is, how the substance meets the criteria for addition and deletion. A short summary of the rationale must be submitted along with a more detailed narrative; and

(g)(6) Supporting data; that is, the petition must include sufficient information to scientifically support the request to modify the list. Such information shall include:

(g)(6)(i) A list of all support documents;

(g)(6)(ii) Documentation of literature searches conducted, including, but not limited to, identification of the database(s) searched, the search strategy, dates covered, and printed results;

(g)(6)(iii) Effects data (animal, human, and environmental test data) indicating the potential for death, injury, or serious adverse human and environmental impacts from acute exposure following an accidental release; printed copies of the data sources, in English, should be provided; and

(g)(6)(iv) Exposure data or previous accident history data, indicating the potential for serious adverse human health or environmental effects from an accidental release. These data may include, but are not limited to, physical and chemical properties of the substance, such as vapor pressure; modeling results, including data and assumptions used and model documentation; and historical accident data, citing data sources.

(h) Within 18 months of receipt of a petition, the Administrator shall publish in the Federal Register a notice either denying the petition or granting the petition and proposing a listing.

68.125 Exemptions

Agricultural nutrients. Ammonia used as an agricultural nutrient, when held by farmers, is exempt from all provisions of this part.

SUBPART G—RISK MANAGEMENT PLAN

68.150 Submission

(a) The owner or operator shall submit a single RMP that includes the information required by §§68.155 through 68.185 for all covered processes. The RMP shall be submitted in a method and format to a central point as specified by EPA prior to June 21, 1999.

(b) The owner or operator shall submit the first RMP no later than the latest of the following dates:

(b)(1) June 21, 1999;

(b)(2) Three years after the date on which a regulated substance is first listed under §68.130; or

(b)(3) The date on which a regulated substance is first present above a threshold quantity in a process.

(c) Subsequent submissions of RMPs shall be in accordance with §68.190.

(d) Notwithstanding the provisions of §§68.155 to 68.190, the RMP shall exclude classified information. Subject to appropriate procedures to protect such information from public disclosure, classified data or information excluded from the RMP may be made available in a classified annex to the RMP for review by Federal and state representatives who have received the appropriate security clearances.

68.155 Executive Summary

The owner or operator shall provide in the RMP an executive summary that includes a brief description of the following elements:

(a) The accidental release prevention and emergency response policies at the stationary source;

(b) The stationary source and regulated substances handled;

(c) The worst-case release scenario(s) and the alternative release scenario(s), including administrative controls and mitigation measures to limit the distances for each reported scenario;

(d) The general accidental release prevention program and chemical-specific prevention steps;

(e) The five-year accident history;

(f) The emergency response program; and

(g) Planned changes to improve safety.

68.160 Registration

(a) The owner or operator shall complete a single registration form and include it in the RMP. The form shall cover all regulated substances handled in covered processes.

(b) The registration shall include the following data:

(b)(1) Stationary source name, street, city, county, state, zip code, latitude, and longitude;

(b)(2) The stationary source Dun and Bradstreet number;

(b)(3) Name and Dun and Bradstreet number of the corporate parent company;

(b)(4) The name, telephone number, and mailing address of the owner or operator;

(b)(5) The name and title of the person or position with overall responsibility for RMP elements and implementation;

(b)(6) The name, title, telephone number, and 24-hour telephone number of the emergency contact;

(b)(7) For each covered process, the name and CAS number of each regulated substance held above the threshold quantity in the process, the maximum quantity of each regulated substance or mixture in the process (in pounds) to two significant digits, the SIC code, and the Program level of the process;

(b)(8) The stationary source EPA identifier;

(b)(9) The number of full-time employees at the stationary source;

(b)(10) Whether the stationary source is subject to 29 CFR 1910.119;

(b)(11) Whether the stationary source is subject to 40 CFR part 355;

(b)(12) Whether the stationary source has a CAA Title V operating permit; and

(b)(13) The date of the last safety inspection of the stationary source by a Federal, state, or local government agency and the identity of the inspecting entity.

68.165 Offsite Consequence Analysis

(a) The owner or operator shall submit in the RMP information:

(a)(1) One worst-case release scenario for each Program 1 process; and

(a)(2) For Program 2 and 3 processes, one worst-case release scenario to represent all regulated toxic substances held above the threshold quantity and one worst-case release scenario to represent all regulated flammable substances held above the threshold quantity. If additional worst-case scenarios for toxics or flammables are required by §68.25(a)(2)(iii), the owner or operator shall submit the same information on the additional scenario(s). The owner or operator of Program 2 and 3 processes shall also submit information on one alternative release scenario for each regulated toxic substance held above the threshold quantity and one alternative release scenario to represent all regulated flammable substances held above the threshold quantity.

(b) The owner or operator shall submit the following data:

(b)(1) Chemical name;

(b)(2) Physical state (toxics only);

(b)(3) Basis of results (give model name if used);

(b)(4) Scenario (explosion, fire, toxic gas release, or liquid spill and vaporization);

(b)(5) Quantity released in pounds;

(b)(6) Release rate;

(b)(7) Release duration;

(b)(8) Wind speed and atmospheric stability class (toxics only);

(b)(9) Topography (toxics only);

(b)(10) Distance to endpoint;

(b)(11) Public and environmental receptors within the distance;

(b)(12) Passive mitigation considered; and

(b)(13) Active mitigation considered (alternative releases only);

68.168 Five-Year Accident History

The owner or operator shall submit in the RMP the information provided in §68.42(b) on each accident covered by §68.42(a).

68.170 Prevention Program/Program 2

(a) For each Program 2 process, the owner or operator shall provide in the RMP the information indicated in paragraphs (b) through (k) of this section. If the same information applies to more than one covered process, the owner or operator may provide the information only once, but shall indicate to which processes the information applies.

(b) The SIC code for the process.

(c) The name(s) of the chemical(s) covered.

(d) The date of the most recent review or revision of the safety information and a list of Federal or state regulations or industry-specific design codes and standards used to demonstrate compliance with the safety information requirement.

(e) The date of completion of the most recent hazard review or update.

(e)(1) The expected date of completion of any changes resulting from the hazard review;

(e)(2) Major hazards identified;

(e)(3) Process controls in use;

(e)(4) Mitigation systems in use;

(e)(5) Monitoring and detection systems in use; and

(e)(6) Changes since the last hazard review.

(f) The date of the most recent review or revision of operating procedures.

(g) The date of the most recent review or revision of training programs;

(g)(1) The type of training provided—classroom, classroom plus on the job, on the job; and

(g)(2) The type of competency testing used.

(h) The date of the most recent review or revision of maintenance procedures and the date of the most recent equipment inspection or test and the equipment inspected or tested.

(i) The date of the most recent compliance audit and the expected date of completion of any changes resulting from the compliance audit.

(j) The date of the most recent incident investigation and the expected date of completion of any changes resulting from the investigation.

(k) The date of the most recent change that triggered a review or revision of safety information, the hazard review, operating or maintenance procedures, or training.

68.175 Prevention Program/Program 3.

(a) For each Program 3 process, the owner or operator shall provide the information indicated in paragraphs (b) through (p) of this section. If the same information applies to more than one covered process, the owner or operator may provide the information only once, but shall indicate to which processes the information applies.

(b) The SIC code for the process.

(c) The name(s) of the substance(s) covered.

(d) The date on which the safety information was last reviewed or revised.

(e) The date of completion of the most recent PHA or update and the technique used.

(e)(1) The expected date of completion of any changes resulting from the PHA;

(e)(2) Major hazards identified;

(e)(3) Process controls in use;

(e)(4) Mitigation systems in use;

(e)(5) Monitoring and detection systems in use; and

(e)(6) Changes since the last PHA.

(f) The date of the most recent review or revision of operating procedures.

(g) The date of the most recent review or revision of training programs;

(g)(1) The type of training provided—classroom, classroom plus on the job, on the job; and

(g)(2) The type of competency testing used.

(h) The date of the most recent review or revision of maintenance procedures and the date of the most recent equipment inspection or test and the equipment inspected or tested.

(i) The date of the most recent change that triggered management of change procedures and the date of the most recent review or revision of management of change procedures.

(j) The date of the most recent pre-startup review.

(k) The date of the most recent compliance audit and the expected date of completion of any changes resulting from the compliance audit;

(l) The date of the most recent incident investigation and the expected date of completion of any changes resulting from the investigation;

(m) The date of the most recent review or revision of employee participation plans;

(n) The date of the most recent review or revision of hot work permit procedures;

(o) The date of the most recent review or revision of contractor safety procedures; and

(p) The date of the most recent evaluation of contractor safety performance.

68.180 Emergency Response Program.

(a) The owner or operator shall provide in the RMP the following information:

(a)(1) Do you have a written emergency response plan?

(a)(2) Does the plan include specific actions to be taken in response to an accidental releases of a regulated substance?

(a)(3) Does the plan include procedures for informing the public and local agencies responsible for responding to accidental releases?

(a)(4) Does the plan include information on emergency health care?

(a)(5) The date of the most recent review or update of the emergency response plan;

(a)(6) The date of the most recent emergency response training for employees.

(b) The owner or operator shall provide the name and telephone number of the local agency with which the plan is coordinated.

(c) The owner or operator shall list other Federal or state emergency plan requirements to which the stationary source is subject.

68.185 Certification

(a) For Program 1 processes, the owner or operator shall submit in the RMP the certification statement provided in §68.12(b)(4).

(b) For all other covered processes, the owner or operator shall submit in the RMP a single certification that, to the best of the signer's knowledge, information, and belief formed after reasonable inquiry, the information submitted is true, accurate, and complete.

68.190 Updates

(a) The owner or operator shall review and update the RMP as specified in paragraph (b) of this section and submit it in a method and format to acentral point specified by EPA prior to June 21, 1999.

(b) The owner or operator of a stationary source shall revise and update the RMP submitted under §68.150 as follows:

(b)(1) Within five years of its initial submission or most recent update required by paragraphs (b)(2) through (b)(7) of this section, whichever is later.

(b)(2) No later than three years after a newly regulated substance is first listed by EPA;

(b)(3) No later than the date on which a new regulated substance is first present in an already covered process above a threshold quantity;

(b)(4) No later than the date on which a regulated substance is first present above a threshold quantity in a new process;

(b)(5) Within six months of a change that requires a revised PHA or hazard review;

(b)(6) Within six months of a change that requires a revised offsite consequence analysis as provided in §68.36; and

(b)(7) Within six months of a change that alters the Program level that applied to any covered process.

(c) If a stationary source is no longer subject to this part, the owner or operator shall submit a revised registration to EPA within six months indicating that the stationary source is no longer covered.

SUBPART H—OTHER REQUIREMENTS

68.200 Recordkeeping

The owner or operator shall maintain records supporting the implementation of this part for five years unless otherwise provided in Subpart D of this part.

68.210 Availability of Information to the Public

(a) The RMP required under subpart G of this part shall be available to the public under 42 U.S.C. 7414(c).

(b) The disclosure of classified information by the Department of Defense or other Federal agencies or contractors of such agencies shall be controlled by applicable laws, regulations, or executive orders concerning the release of classified information.

68.215 Permit Content and Air Permitting Authority or Designated Agency Requirements

(a) These requirements apply to any stationary source subject to this part 68 and parts 70 or 71 of this Chapter. The 40 CFR part 70 or part 71 permit for the stationary source shall contain:

(a)(1) A statement listing this part as an applicable requirement;

(a)(2) Conditions that require the source owner or operator to submit:

(a)(2)(i) A compliance schedule for meeting the requirements of this part by the date provided in §68.10(a) or;

(a)(2)(ii) As part of the compliance certification submitted under 40 CFR 70.6(c)(5), a certification statement that the source is in compliance with all requirements of this part, including the registration and submission of the RMP.

(b) The owner or operator shall submit any additional relevant information requested by the air permitting authority or designated agency.

(c) For 40 CFR part 70 or part 71 permits issued prior to the deadline for registering and submitting the RMP and which do not contain permit conditions described in paragraph (a) of this section, the owner or operator or air permitting authority shall initiate permit revision or reopening according to the procedures of 40 CFR 70.7 or 71.7 to incorporate the terms and conditions consistent with paragraph (a) of this section.

(d) The state may delegate the authority to implement and enforce the requirements of paragraph (e) of this section to a state or local agency or agencies other than the air permitting authority. An up-to-date copy of any delegation instrument shall be maintained by the air permitting authority. The state may enter a written agreement with the Administrator under which EPA will implement and enforce the requirements of paragraph (e) of this section.

(e) The air permitting authority or the agency designated by delegation or agreement under paragraph (d) of this section shall, at a minimum:

(e)(1) Verify that the source owner or operator has registered and submitted an RMP or a revised plan when required by this part;

(e)(2) Verify that the source owner or operator has submitted a source certification or in its absence has submitted a compliance schedule consistent with paragraph (a)(2) of this section;

(e)(3) For some or all of the sources subject to this section, use one or more mechanisms such as, but not limited to, a completeness check, source audits, record reviews, or facility inspections to ensure that permitted sources are in compliance with the requirements of this part; and

(e)(4) Initiate enforcement action based on paragraphs (e)(1) and (e)(2) of this section as appropriate.

68.220 Audits

(a) In addition to inspections for the purpose of regulatory development and enforcement of the Act, the implementing agency shall periodically audit RMPs submitted under subpart G of this part to review the adequacy of such RMPs and require revisions of RMPs when necessary to ensure compliance with subpart G of this part.

(b) The implementing agency shall select stationary sources for audits based on any of the following criteria:

(b)(1) Accident history of the stationary source;

(b)(2) Accident history of other stationary sources in the same industry;

(b)(3) Quantity of regulated substances present at the stationary source;

(b)(4) Location of the stationary source and its proximity to the public and environmental receptors;

(b)(5) The presence of specific regulated substances;

(b)(6) The hazards identified in the RMP; and

(b)(7) A plan providing for neutral, random oversight.

(c) Exemption from audits. A stationary source with a Star or Merit ranking under OSHA's voluntary protection program shall be exempt from audits under paragraph (b)(2) and (b)(7) of this section.

(d) The implementing agency shall have access to the stationary source, supporting documentation, and any area where an accidental release could occur.

(e) Based on the audit, the implementing agency may issue the owner or operator of a stationary source a written preliminary determination of necessary revisions to the stationary source's RMP to ensure that the RMP meets the criteria of subpart G of this part. The preliminary determination shall include an explanation for the basis for the revisions, reflecting industry standards and guidelines (such as AIChE/CCPS guidelines and ASME and API standards) to the extent that such standards and guidelines are applicable, and shall include a timetable for their implementation.

(f) Written response to a preliminary determination.

(f)(1) The owner or operator shall respond in writing to a preliminary determination made in accordance with paragraph (e) of this section. The response shall state the owner or operator will implement the revisions contained in the preliminary determination in accordance with the timetable included in the preliminary determination or shall state that the owner or operator rejects the revisions in whole or in part. For each rejected revision, the owner or operator shall explain the basis for rejecting such revision. Such explanation may include substitute revisions.

(f)(2) The written response under paragraph (f)(1) of this section shall be received by the implementing agency within 90 days of the issue of the preliminary determination or a shorter period of time as the implementing agency specifies in the preliminary determination as necessary to protect public health and the environment. Prior to the written response being due and upon written request from the owner or operator, the implementing agency may provide in writing additional time for the response to be received.

(g) After providing the owner or operator an opportunity to respond under paragraph (f) of this section, the implementing agency may issue the owner or operator a written final determination of necessary revisions to the stationary source's RMP. The final determination may adopt or modify the revisions contained in the preliminary determination under paragraph (e) of this section or may adopt or modify the substitute revisions provided in the response under paragraph (f) of this section. A final determination that adopts a revision rejected by the owner or operator shall include an explanation of the basis for the revision. A final determination that fails to adopt a substitute revision provided under paragraph (f) of this section shall include an explanation of the basis for finding such substitute revision unreasonable.

(h) Thirty days after completion of the actions detailed in the implementation schedule set in the final determination under paragraph (g) of this section, the owner or operator shall be in violation of subpart G of this part and this section unless the owner or operator revises the RMP prepared under subpart G of this part as required by the final determination, and submits the revised RMP as required under §68.150.

(i) The public shall have access to the preliminary determinations, responses, and final determinations under this section in a manner consistent with §68.210.

(j) Nothing in this section shall preclude, limit, or interfere in any way with the authority of EPA or the state to exercise its enforcement, investigatory, and information gathering authorities concerning this part under the Act.

Appendix B

RMP List of Regulated Substances

Introduction

The RMP rule includes a list of toxic substances and a list of flammable substances regulated by the rule. This section provides this list and describes the procedures for petitioning the EPA to revise the list.

Regulatory Text

§68.130 List of Substances

(a) Regulated toxic and flammable substances under section 112(r) of the Clean Air Act are the substances listed in Tables 1, 2, 3, and 4 [of § 68.130]. Threshold quantities for listed toxic and flammable substance are specified in the tables.

(b) The basis for placing toxic and flammable substances on the list of regulated substances are explained in the notes to the list.

Chemical Name	CAS No.	Threshold Quantity (lb)	Basis for Listing
Acrolein [2-Propanal]	107-02-8	5,000	b
Acrylonitrile [2-Propenenitrile]	107-13-1	20,000	b
Acrylyl chloride [2-Propenoyl chloride]	814-68-6	5,000	b
Allyl alcohol [2-Propen-1-ol]	107-18-6	15,000	b
Allylamine [2-Propen-1-amine]	107-11-9	10,000	b
Ammonia (anhydrous)	7664-41-7	10,000	a, b

Chemical Name	CAS No.	Threshold Quantity (lb)	Basis for Listing
Ammonia (conc 20% or greater)	7664-41-7	20,000	a, b
Arsenous trichloride	7784-34-1	15,000	b
Arsine	7784-42-1	1,000	b
Boron trichloride [Borane, trichloro-]	10294-34-5	5,000	b
Boron trifluoride [Borane, trifluoro-]	7637-07-2	5,000	b
Boron triflouride compound with methyl ether (1:1) [boron, triflouro (oxybis metane)]-, T-4-	353-42-4	15,000	b
Bromine	7726-95-6	10,000	a, b
Carbon disulfide	75-15-0	20,000	b
Chlorine	7782-50-5	2,500	a, b
Chlorine dioxide [Chlorine oxide (ClO2)]	10049-04-4	1,000	c
Chloroform [Methane, trichloro-]	67-66-3	20,000	b
Chloromethyl ether [Methane, oxybis[chloro-]	542-88-1	1,000	b
Chloromethyl methyl ether [Methane, chloromethoxy-]	107-30-2	5,000	b
Crotonaldehyde [2-Butenal]	4170-30-3	20,000	b
Crotonaldehyde, (E)- [2-Butenal, (E)-]	123-73-9	20,000	b
Cyanogen chloride	506-77-4	10,000	c
Cyclohexylamine [Cyclohexanamine]	108-91-8	15,000	b
Diborane	19287-45-7	2,500	b
Dimethyldichlorosilane [Silane, dichlorodimethyl-]	75-78-5	5,000	b
1,1-Dimethylhydrazine [Hydrazine, 1,1-dimethyl-]	57-14-7	15,000	b

Chemical Name	CAS No.	Threshold Quantity (lb)	Basis for Listing
Epichlorohydrin [Oxirane, (chloromethyl)-]	106-89-8	20,000	b
Ethylenediamine [1,2-Ethanediamine]	107-15-3	20,000	b
Ethyleneimine [Aziridine]	151-56-4	10,000	b
Ethylene oxide [Oxirane]	75-21-8	10,000	a, b
Fluorine	7782-41-4	1,000	b
Formaldehyde (solution)	50-00-0	15,000	b
Furan	110-00-9	5,000	b
Hydrazine	302-01-2	15,000	b
Hydrochloric acid (conc 30% or greater)	7647-01-0	15,000	d
Hydrocyanic acid	74-90-8	2,500	a, b
Hydrogen chloride (anhydrous) [Hydrochloric acid]	7647-01-0	5,000	a
Hydrogen fluoride/Hydrofluoric acid (conc 50% or greater) [Hydrofluoric acid]	7664-39-3	1,000	a, b
Hydrogen selenide	7783-07-5	500	b
Hydrogen sulfide	7783-06-4	10,000	a, b
Iron, pentacarbonyl- [Iron carbonyl (Fe(CO)5), (TB-5-11)-]	13463-40-6	2,500	b
Isobutyronitrile [Propanenitrile, 2-methyl-]	78-82-0	20,000	b
Isopropyl chloroformate [Carbonochloridic acid, 1-methylethyl ester]	108-23-6	15,000	b
Methacrylontrile [2-Propenenitrile, 2-methyl-]	126-98-7	10,000	b
Methyl chloride [Methane, chloro-]	74-87-3	10,000	a
Methyl chloroformate [Carbonochloridic acid, methylester]	79-22-1	5,000	b

Chemical Name	CAS No.	Threshold Quantity (lb)	Basis for Listing
Methyl hydrazine [Hydrazine, methyl-]	60-34-4	15,000	b
Methyl isocyanate [Methane, isocyanato-]	624-83-9	10,000	a, b
Methyl mercaptan [Methanethiol]	74-93-1	10,000	b
Methyl thiocyanate [Thiocyanic acid, methyl ester]	556-64-9	20,000	b
Methyltrichlorosilane [Silane, trichloromethyl-]	75-79-6	5,000	b
Nickel carbonyl	13463-39-3	1,000	b
Nitric acid (conc 80% or greater)	7697-37-2	15,000	b
Nitric oxide [Nitrogen oxide (NO)]	10102-43-9	10,000	b
Oleum (Fuming Sulfuric acid) [Sulfuric acid, mixture with sulfur trioxide]1	8014-95-7	10,000	e
Peracetic acid [Ethaneperoxoic acid]	79-21-0	10,000	b
Perchloromethylmercaptan [Methanesulfenyl chloride, trichloro-]	594-42-3	10,000	b
Phosgene [Carbonic dichloride]	75-44-5	500	a, b
Phosphine	7803-51-2	5,000	b
Phosphorous oxychloride [Phosphoryl chloride]	10025-87-3	5,000	b
Phosphorous trichloride [Phosphorous trichloride]	7719-12-2	15,000	b
Piperidine	110-89-4	15,000	b
Propionitrile [Propanenitrile]	107-12-0	10,000	b
Propyl chloroformate [Carbonochloridic acid, propylester]	109-61-5	15,000	b
Propyleneimine [Aziridine, 2-methyl-]	75-55-8	10,000	b

Chemical Name	CAS No.	Threshold Quantity (lb)	Basis for Listing
Propylene oxide [Oxirane, methyl-]	75-56-9	10,000	b
Sulfur dioxide (anhydrous)	7446-09-5	5,000	a, b
Sulfur tetrafluoride [Sulfur fluoride (SF4), (T-4)-]	7783-60-0	2,500	b
Sulfur trioxide	7446-11-9	10,000	a, b
Tetramethyllead [Plumbane, tetramethyl-]	75-74-1	10,000	b
Tetranitromethane [Methane, tetranitro-]	509-14-8	10,000	b
Titanium tetrachloride [Titanium chloride (TiCl4) (T-4)-]	7550-45-0	2,500	b
Toluene 2,4-diisocyanate [Benzene, 2,4-diisocyanato-1-methyl-]1	584-84-9	10,000	a
Toluene 2,6-diisocyanate [Benzene, 1,3-diisocyanato-2-methyl-]1	91-08-7	10,000	a
Toluene diisocyanate (unspecified isomer) [Benzene, 1,3-diisocyanatomethyl-]1	26471-62-5	10,000	a
Trimethylchlorosilane [Silane, chlorotrimethyl-]	75-77-4	10,000	b
Vinyl acetate monomer [Acetic acid ethenyl ester]	108-05-4	15,000	b

1 The mixture exemption in §68.115(b)(1) does not apply to the substance

Basis for listing:

a Mandated for listing by Congress

b On EHS, vapor pressure 10 mm Hg or greater

c Toxic gas

d Toxicity of hydrogen chloride, potential to release hydrogen chloride, and history of accidents

e Toxicity of sulfur trioxide and sulfuric acid, potential to release sulfur trioxide, and history of accidents

List of Regulated Flammable Substances

Substance Name	CAS No.	Threshold Quantity (lb)	Basis for Listing
Acetaldehyde	75-07-0	10,000	g
Acetylene [Ethyne]	74-86-2	10,000	f
Bromotrifluorethylene [Ethene, bromotrifluoro-]	598-73-2	10,000	f
1,3-Butadiene	106-99-0	10,000	f
Butane	106-97-8	10,000	f
1-Butene	106-98-9	10,000	f
2-Butene	107-01-7	10,000	f
Butene	25167-67-3	10,000	f
2-Butene-cis	590-18-1	10,000	f
2-Butene-trans [2-Butene, (E)]	624-64-6	10,000	f
Carbon oxysulfide [Carbon oxide sulfide (COS)]	463-58-1	10,000	f
Chlorine monoxide [Chlorine oxide]	7791-21-1	10,000	f
2-Chloropropylene [1-Propene, 2-chloro-]	557-98-2	10,000	g
1-Chloropropylene [1-Propene, 1-chloro-]	590-21-6	10,000	g
Cyanogen [Ethanedinitrile]	460-19-5	10,000	f
Cycolpropane	75-19-4	10,000	f
Dichlorosilane [Silane, dichloro-]	4109-96-0	10,000	f
Difluoroethane [Ethane, 1,1-difluoro-]	75-37-6	10,000	f
Dimethylamine [Methanamine, N-methyl-]	124-40-3	10,000	f
2,2-Dimethylpropane [Propane, 2,2-dimethyl]	463-82-1	10,000	f
Ethane	74-84-0	10,000	f
Ethyl acetylene [1-Butyne]	107-00-6	10,000	f
Ethylamine [Ethanamine]	75-04-7	10,000	f

Substance Name	CAS No.	Threshold Quantity (lb)	Basis for Listing
Ethyl chloride [Ethane, chloro-]	75-00-3	10,000	f
Ethylene [Ethene]	74-85-1	10,000	f
Ethyl ether [Ethane, 1,1′-oxybis-]	60-29-7	10,000	g
Ethyl mercaptan [Ethanethiol]	75-08-1	10,000	g
Ethyl nitrite [Nitrous acid, ethyl ester]	109-95-5	10,000	f
Hydrogen	1333-74-0	10,000	f
Isobutane [Propane, 2-methyl-]	75-28-5	10,000	f
Isopentane [Butane, 2-methyl-]	78-78-4	10,000	g
Isoprene [1,3-Butadiene, 2-methyl-]	78-79-5	10,000	g
Isopropylamine [2-Propanamine]	75-31-0	10,000	g
Isopropyl chloride [Propane, 2-chloro-]	75-29-6	10,000	g
Methane	74-82-8	10,000	f
Methylamine [Methanamine]	74-89-5	10,000	f
3-Methyl-1-butene	563-45-1	10,000	f
2-Methyl-1-butene	563-46-2	10,000	g
Methyl ether [Methane, oxybis-]	115-10-6	10,000	f
Methyl formate [Formic acid, methyl ester]	107-31-3	10,000	g
2-Methylpropene [1-Propene, 2-methyl-]	115-11-7	10,000	f
1,3-Pentadiene	504-60-9	10,000	f
Pentane	109-66-0	10,000	g
1-Pentene	109-67-1	10,000	g
2-Pentene, (E)-	646-04-8	10,000	g
2-Pentene, (Z)-	627-20-3	10,000	g
Propadiene [1,2-Propadiene]	463-49-0	10,000	f
Propane	74-98-6	10,000	f
Propylene [1-Propene]	115-07-1	10,000	f

Substance Name	CAS No.	Threshold Quantity (lb)	Basis for Listing
Propyne [1-Propyne]	74-99-7	10,000	f
Silane	7803-62-5	10,000	f
Tetrafluoroethylene [Ethene, tetrafluoro-]	116-14-3	10,000	f
Tetramethylsilane [Silane, tetramethyl-]	75-76-3	10,000	g
Trichlorosilane [Silane, trichloro-]	10025-78-2	10,000	g
Trifluorochloroethylene [Ethene, chlorotrifluoro-]	79-38-9	10,000	f
Trimethylamine [Methanamine, N,N-dimethyl-]	75-50-3	10,000	f
Vinyl acetylene [1-Buten-3-yne]	689-97-4	10,000	f
Vinyl chloride [Ethene, chloro-]	75-01-4	10,000	a, f
Vinyl ethyl ether [Ethene, ethoxy-]	109-92-2	10,000	g
Vinyl fluoride [Ethene, fluoro-]	75-02-5	10,000	f
Vinylidene chloride [Ethene, 1,1-dichloro-]	75-35-4	10,000	g
Vinylidene fluoride [Ethene, 1,1-difluoro-]	75-38-7	10,000	f
Vinyl methyl ether [Ethene, methoxy-]	107-25-5	10,000	f

Basis for listing:
a Mandated for listing by Congress
f Flammable gas
g Volatile flammable liquid

Changing the Lists

EPA has specified in §68.120 the methodology for adding to or deleting from the lists of regulated toxic and flammable substances. Any person may petition to add or delete substances from the lists. Acceptance or denial of the petition must be published in the *Federal Register* within 18 months of EPA's receipt of the petition. The petition must include a number of elements, including documentation supporting the technical basis of the proposed list change. The requirements are given in the regulatory text.

The criteria for acceptance of a proposed change are described below:

- No substance with a natinal primary ambient air quality standard may be added to the list.
- No substance regulated under Title VI of the CAA may be added to the list.
- A substance may be added to the list if an accidental release of the substance might reasonably be expected to cause serious adverse effects to human health or the environment.
- A substance may be deleted from the list if it can be shown that an accidental release of the substance might not reasonably be expected to cause serious adverse effects to human health or the environment.

Appendix C

Text of the OSHA Process Safety Management (PSM) Standard

Process Safety Management of Highly Hazardous Chemicals

Purpose. This section contains requirements for preventing or minimizing the consequences of catastrophic releases of toxic, reactive, flammable, or explosive chemicals. These releases may result in toxic, fire or explosion hazards.

(a) Application

(a)(1) This section applies to the following:

(a)(1)(i) A process which involves a chemical at or above the specified threshold quantities listed in Appendix A to this section;

(a)(1)(ii) A process which involves a flammable liquid or gas (as defined in 1910.1200(c) of this part) on site in one location, in a quantity of 10,000 pounds (4535.9 kg) or more except for:

(a)(1)(ii)(A) Hydrocarbon fuels used solely for workplace consumption as a fuel (for example, propane used for comfort heating, gasoline for vehicle refueling), if such fuels are not a part of a process containing another highly hazardous chemical covered by this standard;

(a)(1)(ii)(B) Flammable liquids stored in atmospheric tanks or transferred which are kept below their normal boiling point without benefit of chilling or refrigeration.

(a)(2) This section does not apply to:

(a)(2)(i) Retail facilities;

(a)(2)(ii) Oil or gas well drilling or servicing operations; or,

(a)(2)(iii) Normally unoccupied remote facilities.

(b) Definitions

Atmospheric tank means a storage tank which has been designed to operate at pressures from atmospheric through 0.5 p.s.i.g. (pounds per square inch gauge, 3.45 Kpa).

Boiling point means the boiling point of a liquid at a pressure of 14.7 pounds per square inch absolute (p.s.i.a.) (760 mm.). For the purposes of this section, where an accurate boiling point is unavailable for the material in question, or for mixtures which do not have a constant boiling point, the 10 percent point of a distillation performed in accordance with the Standard Method of Test for Distillation of Petroleum Products, ASTM D-86-62, which is incorporated by reference as specified in §1910.6, may be used as the boiling point of the liquid.

Catastrophic release means a major uncontrolled emission, fire, or explosion, involving one or more highly hazardous chemicals, that presents serious danger to employees in the workplace.

Facility means the buildings, containers or equipment which contain a process.

Highly hazardous chemical means a substance possessing toxic, reactive, flammable, or explosive properties and specified by paragraph (a)(1) of this section.

Hot work means work involving electric or gas welding, cutting, brazing, or similar flame or spark-producing operations.

Normally unoccupied remote facility means a facility which is operated, maintained or serviced by employees who visit the facility only periodically to check its operation and to perform necessary operating or maintenance tasks. No employees are permanently stationed at the facility. Facilities meeting this definition are not contiguous with, and must be geographically remote from all other buildings, processes or persons.

Process means any activity involving a highly hazardous chemical including any use, storage, manufacturing, handling, or the on-site movement of such chemicals, or combination of these activities. For purposes of this definition, any group of vessels which are interconnected and separate vessels which are located such that a highly hazardous chemical could be involved in a potential release shall be considered a single process.

Replacement in kind means a replacement which satisfies the design specification.

Trade secret means any confidential formula, pattern, process, device, information or compilation of information that is used in an employer's business, and that gives the employer an opportunity to obtain an advantage over competitors who do not know or use it. Appendix D contained in §1910.1200 sets out the criteria to be used in evaluating trade secrets.

(c) Employee Participation

(c)(1) Employers shall develop a written plan of action regarding the implementation of the employee participation required by this paragraph.

(c)(2) Employers shall consult with employees and their representatives on the conduct and development of process hazards analyses and on the development of the other elements of process safety management in this standard.

(c)(3) Employers shall provide to employees and their representatives access to process hazard analyses and to all other information required to be developed under this standard.

(d) Process Safety Information

In accordance with the schedule set forth in paragraph (e)(1) of this section, the employer shall complete a compilation of written process safety information before conducting any process hazard analysis required by the standard. The compilation of written process safety information is to enable the employer and the employees involved in operating the process to identify and understand the hazards posed by those processes involving highly hazardous chemicals. This process safety information shall include information pertaining to the hazards of the highly hazardous chemicals used or produced by the process, information pertaining to the technology of the process, and information pertaining to the equipment in the process.

(d)(1) Information pertaining to the hazards of the highly hazardous chemicals in the process. This information shall consist of at least the following:

(d)(1)(i) Toxicity information;

(d)(1)(ii) Permissible exposure limits;

(d)(1)(iii) Physical data;

(d)(1)(iv) Reactivity data:

(d)(1)(v) Corrosivity data;

(d)(1)(vi) Thermal and chemical stability data; and

(d)(1)(vii) Hazardous effects of inadvertent mixing of different materials that could foreseeably occur. Note: Material Safety Data Sheets meeting the requirements of 29 CFR 1910.1200(g) may be used to comply with this requirement to the extent they contain the information required by this subparagraph.

(d)(2) Information pertaining to the technology of the process.

(d)(2)(i) Information concerning the technology of the process shall include at least the following:

(d)(2)(i)(A) A block flow diagram or simplified process flow diagram (see Appendix B to this section);

(d)(2)(i)(B) Process chemistry;

(d)(2)(i)(C) Maximum intended inventory;

(d)(2)(i)(D) Safe upper and lower limits for such items as temperatures, pressures, flows or compositions; and,

(d)(2)(i)(E) An evaluation of the consequences of deviations, including those affecting the safety and health of employees.

(d)(2)(ii) Where the original technical information no longer exists, such information may be developed in conjunction with the process hazard analysis in sufficient detail to support the analysis.

(d)(3) Information pertaining to the equipment in the process.

(d)(3)(i)Information pertaining to the equipment in the process shall include:

(d)(3)(i)(A) Materials of construction;

(d)(3)(i)(B) Piping and instrument diagrams (P&ID's);

(d)(3)(i)(C) Electrical classification;

(d)(3)(i)(D) Relief system design and design basis;

(d)(3)(i)(E) Ventilation system design;

(d)(3)(i)(F) Design codes and standards employed;

(d)(3)(i)(G) Material and energy balances for processes built after May 26, 1992; and, (d)(3)(i)(H) Safety systems (for example interlocks, detection or suppression systems).

(d)(3)(ii) The employer shall document that equipment complies with recognized and generally accepted good engineering practices.

(d)(3)(iii) For existing equipment designed and constructed in accordance with codes, standards, or practices that are no longer in general use, the employer shall determine and document that the equipment is designed, maintained, inspected, tested, and operating in a safe manner.

(e) Process Hazard Analysis

(e)(1) The employer shall perform an initial process hazard analysis (hazard evaluation) on processes covered by this standard. The process hazard analysis shall be appropriate to the complexity of the process and shall identify, evaluate, and control the hazards involved in the process. Employers shall determine and document the priority order for conducting process hazard analyses based on a rationale which includes such considerations as extent of the process hazards, number of potentially affected employees, age of the process, and operating history of the process. The process hazard analysis shall be conducted as soon as possible, but not later than the following schedule:

(e)(1)(i) No less than 25 percent of the initial process hazards analyses shall be completed by May 26, 1994;

(e)(1)(ii) No less than 50 percent of the initial process hazards analyses shall be completed by May 26, 1995;

(e)(1)(iii) No less than 75 percent of the initial process hazards analyses shall be completed by May 26, 1996;

(e)(1)(iv) All initial process hazards analyses shall be completed by May 26, 1997.

(e)(1)(v) Process hazards analyses completed after May 26, 1987 which meet the requirements of this paragraph are acceptable as initial process hazards analyses. These process hazard analyses shall be updated and revalidated, based on their completion date, in accordance with paragraph (e)(6) of this section.

(e)(2) The employer shall use one or more of the following methodologies that are appropriate to determine and evaluate the hazards of the process being analyzed.

(e)(2)(i) What-If;

(e)(2)(ii) Checklist;

(e)(2)(iii) What-If/Checklist;

(e)(2)(iv) Hazard and Operability Study (HAZOP):

(e)(2)(v) Failure Mode and Effects Analysis (FMEA);

(e)(2)(vi) Fault Tree Analysis; or

(e)(2)(vii) An appropriate equivalent methodology.

(e)(3) The process hazard analysis shall address:

(e)(3)(i) The hazards of the process;

(e)(3)(ii) The identification of any previous incident which had a likely potential for catastrophic consequences in the workplace;

(e)(3)(iii) Engineering and administrative controls applicable to the hazards and their interrelationships such as appropriate application of detection methodologies to provide early warning of releases. (Acceptable detection methods might include process monitoring and control instrumentation with alarms, and detection hardware such as hydrocarbon sensors.);

(e)(3)(iv) Consequences of failure of engineering and administrative controls;

(e)(3)(v) Facility siting;

(e)(3)(vi) Human factors; and

(e)(3)(vii) A qualitative evaluation of a range of the possible safety and health effects of failure of controls on employees in the workplace.

(e)(4) The process hazard analysis shall be performed by a team with expertise in engineering and process operations, and the team shall include at least one employee who has experience and knowledge specific to the process being evaluated. Also, one member of the team must be knowledgeable in the specific process hazard analysis methodology being used.

(e)(5) The employer shall establish a system to promptly address the team's findings and recommendations; assure that the recommendations are resolved in a timely manner and that the resolution is documented; document what actions are to be taken; complete actions as soon as possible; develop a written schedule of when these actions are to be completed; communicate the actions to operating, maintenance and other employees whose work assignments are in the process and who may be affected by the recommendations or actions.

(e)(6) At least every five (5) years after the completion of the initial process hazard analysis, the process hazard analysis shall be updated and revalidated by a team meeting the requirements in paragraph (e)(4) of this section, to assure that the process hazard analysis is consistent with the current process.

(e)(7) Employers shall retain process hazards analyses and updates or revalidations for each process covered by this section, as well as the documented resolution of recommendations described in paragraph (e)(5) of this section for the life of the process.

(f) Operating Procedures

(f)(1) The employer shall develop and implement written operating procedures that provide clear instructions for safely conducting activities involved in each covered process consistent with the process safety information and shall address at least the following elements.

(f)(1)(i) Steps for each operating phase:

(f)(1)(i)(A) Initial startup;

(f)(1)(i)(B) Normal operations;

(f)(1)(i)(C) Temporary operations;

(f)(1)(i)(D) Emergency shutdown including the conditions under which emergency shutdown is required, and the assignment of shutdown responsibility to qualified operators to ensure that emergency shutdown is executed in a safe and timely manner.

(f)(1)(i)(E) Emergency Operations;

(f)(1)(i)(F) Normal shutdown; and,

(f)(1)(i)(G) Startup following a turnaround, or after an emergency shutdown.

(f)(1)(ii) Operating limits:

(f)(1)(ii)(A) Consequences of deviation; and

(f)(1)(ii)(B) Steps required to correct or avoid deviation.

(f)(1)(iii) Safety and health considerations:

(f)(1)(iii)(A) Properties of, and hazards presented by, the chemicals used in the process;

(f)(1)(iii)(B) Precautions necessary to prevent exposure, including engineering controls, administrative controls, and personal protective equipment;

(f)(1)(iii)(C) Control measures to be taken if physical contact or airborne exposure occurs;

(f)(1)(iii)(D) Quality control for raw materials and control of hazardous chemical inventory levels; and,

(f)(1)(iii)(E) Any special or unique hazards.

(f)(1)(iv) Safety systems and their functions.

(f)(2) Operating procedures shall be readily accessible to employees who work in or maintain a process.

(f)(3) The operating procedures shall be reviewed as often as necessary to assure that they reflect current operating practice, including changes that result from changes in process chemicals, technology, and equipment, and changes to facilities. The employer shall certify annually that these operating procedures are current and accurate.

(f)(4) The employer shall develop and implement safe work practices to provide for the control of hazards during operations such as lockout/tagout; confined space entry; opening process equipment or piping; and control over entrance into a facility by maintenance, contractor, laboratory, or other support personnel. These safe work practices shall apply to employees and contractor employees.

(g) Training

(g)(1)*Initial training.*

(g)(1)(i) Each employee presently involved in operating a process, and each employee before being involved in operating a newly assigned process, shall be trained in an overview of the process and in the operating procedures as specified in paragraph (f) of this section. The training shall include emphasis on the specific safety and health hazards, emergency operations including shutdown, and safe work practices applicable to the employee's job tasks.

(g)(1)(ii) In lieu of initial training for those employees already involved in operating a process on May 26, 1992, an employer may certify in writing that the employee has the required knowledge, skills, and abilities to safely carry out the duties and responsibilities as specified in the operating procedures.

(g)(2) *Refresher training.* Refresher training shall be provided at least every three years, and more often if necessary, to each employee involved in operating a process to assure that the employee understands and adheres to the current operating procedures of the process.

The employer, in consultation with the employees involved in operating the process, shall determine the appropriate frequency of refresher training.

(g)(3) *Training documentation.* The employer shall ascertain that each employee involved in operating a process has received and understood the training required by this paragraph. The employer shall prepare a record which contains the identity of the employee, the date of training, and the means used to verify that the employee understood the training.

(h) Contractors

(h)(1) *Application.* This paragraph applies to contractors performing maintenance or repair, turnaround, major renovation, or specialty work on or adjacent to a covered process. It does not apply to contractors providing incidental services which do not influence process safety, such as janitorial work, food and drink services, laundry, delivery or other supply services.

(h)(2) *Employer responsibilities.*

(h)(2)(i) The employer, when selecting a contractor, shall obtain and evaluate information regarding the contract employer's safety performance and programs.

(h)(2)(ii) The employer shall inform contract employers of the known potential fire, explosion, or toxic release hazards related to the contractor's work and the process.

(h)(2)(iii) The employer shall explain to contract employers the applicable provisions of the emergency action plan required by paragraph (n) of this section.

(h)(2)(iv) The employer shall develop and implement safe work practices consistent with paragraph (f)(4) of this section, to control the entrance, presence and exit of contract employers and contract employees in covered process areas.

(h)(2)(v) The employer shall periodically evaluate the performance of contract employers in fulfilling their obligations as specified in paragraph (h)(3) of this section.

(h)(2)(vi) The employer shall maintain a contract employee injury and illness log related to the contractor's work in process areas.

(h)(3) Contract employer responsibilities. (i) The contract employer shall assure that each contract employee is trained in the work practices necessary to safely perform his/her job.

(h)(3)(ii) The contract employer shall assure that each contract employee is instructed in the known potential fire, explosion, or toxic release hazards related to his/her job and the process, and the applicable provisions of the emergency action plan.

(h)(3)(iii) The contract employer shall document that each contract employee has received and understood the training required by this paragraph. The contract employer shall prepare a record which contains the identity of the contract employee, the date of training, and the means used to verify that the employee understood the training.

(h)(3)(iv) The contract employer shall assure that each contract employee follows the safety rules of the facility including the safe work practices required by paragraph (f)(4) of this section.

(h)(3)(v) The contract employer shall advise the employer of any unique hazards presented by the contract employer's work, or of any hazards found by the contract employer's work.

(i) Pre-Startup Safety Review

(i)(1) The employer shall perform a pre-startup safety review for new facilities and for modified facilities when the modification is significant enough to require a change in the process safety information.

(i)(2) The pre-startup safety review shall confirm that prior to the introduction of highly hazardous chemicals to a process:

(i)(2)(i) Construction and equipment is in accordance with design specifications;

(i)(2)(ii) Safety, operating, maintenance, and emergency procedures are in place and are adequate;

(i)(2)(iii) For new facilities, a process hazard analysis has been performed and recommendations have been resolved or implemented before startup; and modified facilities meet the requirements contained in management of change, paragraph (l).

(i)(2)(iv) Training of each employee involved in operating a process has been completed.

(j) Mechanical Integrity

(j)(1) *Application.* Paragraphs (j)(2) through (j)(6) of this section apply to the following process equipment:

(j)(1)(i) Pressure vessels and storage tanks;

(j)(1)(ii) Piping systems (including piping components such as valves);

(j)(1)(iii) Relief and vent systems and devices;

(j)(1)(iv) Emergency shutdown systems;

(j)(1)(v) Controls (including monitoring devices and sensors, alarms, and interlocks) and,

(j)(1)(vi) Pumps.

(j)(2) *Written Procedures.* The employer shall establish and implement written procedures to maintain the on-going integrity of process equipment.

(j)(3) *Training for process maintenance activities.* The employer shall train each employee involved in maintaining the on-going integrity of process equipment in an overview of that process and its hazards and in the procedures applicable to the employee's job tasks to assure that the employee can perform the job tasks in a safe manner.

(j)(4) *Inspection and testing.*

(j)(4)(i) Inspections and tests shall be performed on process equipment.

(j)(4)(ii) Inspection and testing procedures shall follow recognized and generally accepted good engineering practices.

(j)(4)(iii) The frequency of inspections and tests of process equipment shall be consistent with applicable manufacturers' recommendations and good engineering practices, and more frequently if determined to be necessary by prior operating experience.

(j)(4)(iv) The employer shall document each inspection and test that has been performed on process equipment. The documentation shall identify the date of the inspection or test, the name of the person who performed the inspection or test, the serial number or other identifier of the equipment on which the inspection or test was performed, a description of the inspection or test performed, and the results of the inspection or test.

(j)(5) *Equipment deficiencies.* The employer shall correct deficiencies in equipment that are outside acceptable limits (defined by the process safety information in paragraph (d) of this section) before further use or in a safe and timely manner when necessary means are taken to assure safe operation.

(j)(6) *Quality assurance.*

(j)(6)(i) In the construction of new plants and equipment, the employer shall assure that equipment as it is fabricated is suitable for the process application for which they will be used.

(j)(6)(ii) Appropriate checks and inspections shall be performed to assure that equipment is installed properly and consistent with design specifications and the manufacturer's instructions.

(j)(6)(iii) The employer shall assure that maintenance materials, spare parts and equipment are suitable for the process application for which they will be used.

(k) Hot Work Permit

(k)(1) The employer shall issue a hot work permit for hot work operations conducted on or near a covered process.

(k)(2) The permit shall document that the fire prevention and protection requirements in 29 CFR 1910.252(a) have been implemented prior to beginning the hot work operations; it shall indicate the date(s) authorized for hot work; and identify the object on which hot work is to be performed. The permit shall be kept on file until completion of the hot work operations.

(l) Management of Change

(l)(1) The employer shall establish and implement written procedures to manage changes (except for "replacements in kind") to process chemicals, technology, equipment, and procedures; and, changes to facilities that affect a covered process.

(l)(2) The procedures shall assure that the following considerations are addressed prior to any change:

(l)(2)(i) The technical basis for the proposed change;

(l)(2)(ii) Impact of change on safety and health;

(l)(2)(iii) Modifications to operating procedures;

(l)(2)(iv) Necessary time period for the change; and,

(l)(2)(v) Authorization requirements for the proposed change.

(l)(3) Employees involved in operating a process and maintenance and contract employees whose job tasks will be affected by a change in the process shall be informed of, and trained in, the change prior to start-up of the process or affected part of the process.

(l)(4) If a change covered by this paragraph results in a change in the process safety information required by paragraph (d) of this section, such information shall be updated accordingly.

(l)(5) If a change covered by this paragraph results in a change in the operating procedures or practices required by paragraph (f) of this section, such procedures or practices shall be updated accordingly.

(m) Incident Investigation

(m)(1) The employer shall investigate each incident which resulted in, or could reasonably have resulted in a catastrophic release of highly hazardous chemical in the workplace.

(m)(2) An incident investigation shall be initiated as promptly as possible, but not later than 48 hours following the incident.

(m)(3) An incident investigation team shall be established and consist of at least one person knowledgeable in the process involved, including a contract employee if the incident involved work of the contractor, and other persons with appropriate knowledge and experience to thoroughly investigate and analyze the incident.

(m)(4) A report shall be prepared at the conclusion of the investigation which includes at a minimum:

(m)(4)(i) Date of incident;

(m)(4)(ii) Date investigation began;

(m)(4)(iii) A description of the incident;

(m)(4)(iv) The factors that contributed to the incident; and,

(m)(4)(v) Any recommendations resulting from the investigation.

(m)(5) The employer shall establish a system to promptly address and resolve the incident report findings and recommendations. Resolutions and corrective actions shall be documented.

(m)(6) The report shall be reviewed with all affected personnel whose job tasks are relevant to the incident findings including contract employees where applicable.

(m)(7) Incident investigation reports shall be retained for five years.

(n) Emergency Planning and Response

The employer shall establish and implement an emergency action plan for the entire plant in accordance with the provisions of 29 CFR 1910.38(a). In addition, the emergency action plan shall include procedures for handling small releases. Employers covered under this standard may also be subject to the hazardous waste and emergency response provisions contained in 29 CFR 1910.120(a), (p) and (q).

(o) Compliance Audits

(o)(1) Employers shall certify that they have evaluated compliance with the provisions of this section at least every three years to verify that the procedures and practices developed under the standard are adequate and are being followed.

(o)(2) The compliance audit shall be conducted by at least one person knowledgeable in the process.

(o)(3) A report of the findings of the audit shall be developed.

(o)(4) The employer shall promptly determine and document an appropriate response to each of the findings of the compliance audit, and document that deficiencies have been corrected.

(o)(5) Employers shall retain the two (2) most recent compliance audit reports.

(p) Trade Secrets

(p)(1) Employers shall make all information necessary to comply with the section available to those persons responsible for compiling the process safety information (required by paragraph (d) of this section), those assisting in the development of the process hazard analysis (required by paragraph (e) of this section), those responsible for developing the operating procedures (required by paragraph (f) of this section), and those involved in incident investigations (required by paragraph (m) of this section), emergency planning and response (paragraph (n) of this section) and compliance audits (paragraph (o) of this section) without regard to possible trade secret status of such information.

(p)(2) Nothing in this paragraph shall preclude the employer from requiring the persons to whom the information is made available under paragraph (p)(1) of this section to enter into confidentiality agreements not to disclose the information as set forth in 29 CFR 1910.1200.

(p)(3) Subject to the rules and procedures set forth in 29 CFR 1910.1200(i)(1) through 1910.1200(i)(12), employees and their designated representatives shall have access to trade secret information contained within the process hazard analysis and other documents required to be developed by this standard.

Appendix D

Comparison of OSHA and EPA Lists of Highly Hazardous Chemicals and Regulated Substances

CHEMICAL		LISTING BASIS (See Notes)		THRESHOLD QUANTITY, pounds	
CAS #	NAME	OSHA	EPA	OSHA TQ [1]	EPA TQ [2]
TOXICS (OSHA AND EPA) AND REACTIVES (OSHA)					
75-07-0	Acetaldehyde	T	See flammables	2,500	
107-02-8	Acrolein [2-Propenal]	T,R	(b)	150	5,000
107-13-1	Acrylonitrile [2-Propenenitrile]	(6)	(b)		20,000
814-68-6	Acrylyl chloride [2-Propenoyl chloride]	T	(b)	250	5,000
Various	Alkylaluminums	R		5,000	
107-18-6	Allyl alcohol [2-Propen-1-ol]	(6)	(b)		15,000
107-05-1	Allyl chloride	T		1,000	
107-11-9	Allylamine [2-Propen-1-amine]	T	(b)	1,000	10,000
7664-41-7	Ammonia [anhydrous]	T	(a,b)	10,000	10,000 [3]
7664-41-7	Ammonia solutions (44% ammonia or greater by wt)	T		15,000	

CHEMICAL		LISTING BASIS (See Notes)		THRESHOLD QUANTITY, pounds	
CAS #	NAME	OSHA	EPA	OSHA TQ [1]	EPA TQ [2]
7664-41-7	Ammonia solutions (20% or greater by wt)		(a,b)		20,000[3]
7790-98-9	Ammonium perchlorate	R		7500	
7787-36-2	Ammonium permanganate	R		7500	
7784-34-1	Arsenous trichloride		(b)		15,000
7784-42-1	Arsine [Arsenic hydride]	T	(b)	100	1000
10294-34-5	Boron trichloride [Borane, trichloro-]	T	(b)	2500	5000
7637-07-2	Boron trifluoride [Borane, trifluoro-]	T	(b)	250	5000
353-42-4	Boron trifluoride with Methyl ether (1:1) [Boron, trifluoro [oxybis[metane]]-, T-4-		(b)		15,000
7726-95-6	Bromine	T	(a,b)	1500	10,000
13863-41-7	Bromine chloride	T		1500	
7789-30-2	Bromine pentafluoride	T		2500	
7787-71-5	Bromine trifluoride	R		15,000	
106-96-7	Bromopropyne,3- [Propargyl bromide]	R		100	
75-91-2	Butyl hydroperoxide (tertiary)	R		5,000	
614-45-9	Butyl perbenzoate (tertiary)	R		7,500	
75-15-0	Carbon disulfide	[6]	(b)		20,000
353-50-4	Carbonyl fluoride	T		2,500	
9004-70-0	Cellulose nitrate (conc. > 12.6% nitrogen)	R		2,500	
7782-50-5	Chlorine	T	(a,b)	1,500	2500

CHEMICAL		LISTING BASIS (See Notes)		THRESHOLD QUANTITY, pounds	
CAS #	NAME	OSHA	EPA	OSHA TQ [1]	EPA TQ [2]
10049-04-4	Chlorine dioxide [Chlorine oxide (ClO_2)]	T	(c)	1000	1000
13637-63-3	Chlorine pentafluoride	T		1000	
7790-91-2	Chlorine trifluoride	T,R		1000	
67-66-3	Chloroform [Methane, trichloro-]	[6]	(b)		20,000
542-88-1	Chloromethyl ether [Methane, oxybis[chloro-]]	T	(b)	100	1000
107-30-2	Chloromethyl methyl ether [Methane, chloromethoxy-]	T	(b)	500	5000
76-06-2	Chloropicrin	T,R		500	
NONE	Chloropicrin and methyl bromide mixture	T,R		1500	
NONE	Chloropicrin and methyl chloride mixture	T,R		1500	
4170-30-3	Crotonaldehyde [2-Butenal]	[6]	(b)		20,000
123-73-9	Crotonaldehyde, (E)- [2-butenal, (E)-]	[6]	(b)		20,000
80-15-9	Cumene hydroperoxide	R		5000	
460-19-5	Cyanogen	T	See flammables	2500	
506-77-4	Cyanogen chloride	T	(c)	500	10,000
675-14-9	Cyanuric fluoride	T		100	
108-91-8	Cyclohexylamine [Cyclohexanamine]	[6]	(b)		15,000
110-22-5	Diacetyl peroxide (conc. > 70%)	R		5000	

CHEMICAL		LISTING BASIS (See Notes)		THRESHOLD QUANTITY, pounds	
CAS #	NAME	OSHA	EPA	OSHA TQ [1]	EPA TQ [2]
334-88-3	Diazomethane	T		500	
94-36-0	Dibenzoyl peroxide	R		7500	
19287-45-7	Diborane	T,R	(b)	100	2500
110-05-4	Dibutyl peroxide (tertiary)	R		5000	
7572-29-4	Dichloroacetylene	T		250	
4109-96-0	Dichlorosilane	T	See flam-mables	2500	
557-20-0	Diethyl zinc	R		10,000	
96-10-6	Diethylaluminum chloride [chlorodiethylaluminum]	R		5000	
105-64-6	Diisopropyl peroxydicarbonate	R		7500	
105-74-8	Dilauroyl peroxide	R		7500	
124-40-3	Dimethylamine [anhydrous]	T	See flamm ables	2500	
75-78-5	Dimethyldichlorosilane [Silane, dichlorodimethyl-]	T	(b)	1000	5000
57-14-7	Dimethylhydrazine, 1,1- [Hydrazine, 1,1-dimethyl-]	T	(b)	1000	15,000
97-02-9	Dinitroaniline,2,4-	R		5000	
97-00-7	Dinitrochlorobenzene [1-chloro-2,4-dinitroben zene]	R		5000	
106-89-8	Epichlorohydrin [Oxirane, (chloromethyl)-]	[6]	(b)		20,000

CHEMICAL		LISTING BASIS (See Notes)		THRESHOLD QUANTITY, pounds	
CAS #	NAME	OSHA	EPA	OSHA TQ [1]	EPA TQ [2]
109-95-5	Ethyl nitrite	T,R	See flammables	5000	
75-04-7	Ethylamine	T	See flammables	7500	
371-62-0	Ethylene fluorohydrin	T		100	
75-21-8	Ethylene oxide [Oxirane]	T,R	(a,b)	5000	10,000
107-15-3	Ethylenediamine [1,2-Ethanediamine]	[6]	(b)		20,000
151-56-4	Ethyleneimine [Aziridine]	T,R	(b)	1000	10,000
7782-41-4	Fluorine	T,R	(b)	1000	1000
50-00-0	Formaldehyde (solution, concentration not specified)		(b)		15,000
50-00-0	Formaldehyde [Formalin] (37% or greater by wt, per 7/28/92 OSHA interpretation letter)	T		1000	
110-00-9	Furan	T	(b)	500	5000
684-16-2	Hexafluoroacetone	T		5000	
302-01-2	Hydrazine		(b)		15,000
7647-01-0	Hydrochloric acid (37% or greater by wt) [7]		(d)		15,000[3]
74-90-8	Hydrocyanic acid [Hydrogen cyanide (anhydrous)]	T	(a,b)	1000	2500
10035-10-6	Hydrogen bromide (anhydrous)	T		5000	

CHEMICAL		LISTING BASIS (See Notes)		THRESHOLD QUANTITY, pounds	
CAS #	NAME	OSHA	EPA	OSHA TQ [1]	EPA TQ [2]
7647-01-0	Hydrogen chloride [Hydrochloric acid, anhydrous]	T	(a)	5000	5000[3]
7664-39-3	Hydrogen fluoride/ Hydrofluoric acid (50% or greater by weight)		(a,b)		1000
7664-39-3	Hydrogen fluoride [Hydrofluoric acid (anhydrous)]	T		1000	
7722-84-1	Hydrogen peroxide (52% or greater by wt)	R		7500	
7783-07-5	Hydrogen selenide	T	(b)	150	500
7783-06-4	Hydrogen sulfide	T	(a,b)	1500	10,000
7803-49-8	Hydroxylamine	R		2500	
13463-40-6	Iron, pentacarbonyl- [[Iron carbonyl ($Fe(CO)_5$), (TB-5-11)-]	T	(b)	250	2500
78-82-0	Isobutyronitrile [Propanenitrile, 2-methyl-]	[6]	(b)		20,000
108-23-6	Isopropyl chloroformate [Carbonochloridic acid, methylethyl ester]		(b)		15,000
75-31-0	Isopropylamine	T	See flammables	5000	
463-51-4	Ketene	T		100	
78-85-3	Methacrylaldehyde	T		1000	
126-98-7	Methacrylonitrile [2-Propenenitrile, 2-methyl-]	T	(b)	250	10,000
920-46-7	Methacryloyl chloride	T		150	

CHEMICAL		LISTING BASIS (See Notes)		THRESHOLD QUANTITY, pounds	
CAS #	NAME	OSHA	EPA	OSHA TQ [1]	EPA TQ [2]
30674-80-7	Methacryloyloxyethyl isocyanate	T		100	
74-83-9	Methyl bromide	T		2500	
74-87-3	Methyl chloride [methane, chloro-]	T	(a)	15,000	10,000[4]
79-22-1	Methyl chloroformate [Carbonochloridic acid, methyl ester]	T	(b)	500	5000
1338-23-4	Methyl ethyl ketone peroxide (60% or greater by wt)	R		5000	
453-18-9	Methyl fluoroacetate	T		100	
421-20-5	Methyl fluorosulfate	T		100	
60-34-4	Methyl hydrazine [Hydrazine, methyl-]	T	(b)	100	15,000
74-88-4	Methyl iodide	T		7500	
624-83-9	Methyl isocyanate [Methane, -isocyanato-]	T	(a,b)	250	10,000
74-93-1	Methyl mercaptan [Methanethiol]	T	(b)	5000	10,000
556-64-9	Methyl thiocyanate [Thyocyanic acid, methyl ester]		(b)		20,000
79-84-4	Methyl vinyl ketone	T		100	
74-89-5	Methylamine (anhydrous)	T	See flam-mables	1000	
75-79-6	Methyltrichlorosilane [Silane, trichloromethyl-]	T	(b)	500	5000
13463-39-3	Nickel carbonyl [Nickel tetracarbonyl]	T,R	(b)	150	1000

CHEMICAL		LISTING BASIS (See Notes)		THRESHOLD QUANTITY, pounds	
CAS #	NAME	OSHA	EPA	OSHA TQ [1]	EPA TQ [2]
7697-37-2	Nitric acid (80 % or greater by wt)		(b)		15,000
7697-37-2	Nitric acid (94.5 % or greater by wt)	T		500	
10102-43-9	Nitric oxide [Nitrogen oxide (NO)]	T	(b)	250	10,000
100-01-6	Nitroaniline (Paranitroaniline)	R		5000	
10102-44-0	Nitrogen oxides(NO_2, N_2O_4, N2O3)	T		250	
10544-72-6	Nitrogen tetroxide (Nitrogen peroxide)	T		250	
7783-54-2	Nitrogen trifluoride	T		5000	
10544-73-7	Nitrogen trioxide	T		250	
75-52-5	Nitromethane	R		2500	
8014-95-7	Oleum (65% SO_3 or greater by wt, per 6/24/93OSHA interpretation letter)	T		1000	
8014-95-7	Oleum (Fuming sulfuric acid) [Sulfuric acid, mixture with sulfur trioxide]		(e) [5]		10,000[2]
20816-12-0	Osmium tetroxide	T		100	
7783-41-7	Oxygen difluoride [Fluorine monoxide]	T		100	
10028-15-6	Ozone	T		100	
19624-22-7	Pentaborane	T		100	
79-21-0	Peracetic acid (60% or greater by wt) [Peroxyacetic acid]	T,R		1000	

CHEMICAL		LISTING BASIS (See Notes)		THRESHOLD QUANTITY, pounds	
CAS #	NAME	OSHA	EPA	OSHA TQ [1]	EPA TQ [2]
79-21-0	Peracetic acid (concentration not specified) [Ethaneperoxoic acid]		(b)		10,000
7601-90-3	Perchloric acid (60% or greater by wt)	R		5000	
594-42-3	Perchloromethylmercaptan [Methanesulfenyl chloride, thrichloro-]	T	(b)	150	10,000
7616-94-6	Perchloryl fluoride	T		5000	
75-44-5	Phosgene [carbonic dichloride, carbonyl chloride]	T	(a,b)	100	500
7803-51-2	Phosphine [Hydrogen phosphide]	T	(b)	100	5000
10025-87-3	Phosphorus oxychloride [Phosphoryl chloride]	T	(b)	1000	5000
7719-12-2	Phosphorus trichloride	T	(b)	1000	15,000
110-89-4	Piperidine	(6)	(b)		15,000
107-12-0	Propionitrile (propanenitrile]	(6)	(b)		10,000
109-61-5	Propyl chloroformate [Carbonochloridic acid, propylester]	(6)	(b)		15,000
627-13-4	Propyl nitrate	R		2500	
75-56-9	Propylene oxide [Oxirane, methyl-]	(6)	(b)		10,000
75-55-8	Propyleneimine [Aziridine, 2-methyl-]	(6)	(b)		10,000
107-44-8	Sarin	T		100	
7783-79-1	Selenium hexafluoride	T		1000	
7803-52-3	Stibine [Antimony hydride]	T		500	

CHEMICAL		LISTING BASIS (See Notes)		THRESHOLD QUANTITY, pounds	
CAS #	NAME	OSHA	EPA	OSHA TQ [1]	EPA TQ [2]
7446-09-5	Sulfur dioxide (anhydrous)	T	(a,b)	1000	5000
5714-22-7	Sulfur pentafluoride	T		250	
7783-60-0	Sulfur tetrafluoride [Sulfur fluoride (SF_4), (T-4)-]	T	(b)	250	2500
7446-11-9	Sulfur trioxide [Sulfuric anhydride]	T	(a,b)	1000	10,000
7783-80-4	Tellurium hexafluoride	T		250	
116-14-3	Tetrafluoroethylene	T,R	See flam-mables	5000	
10036-47-2	Tetrafluorohydrazine	T		5000	
75-74-1	Tetramethyl lead [Plumbane, tetramethyl-]	T	(b)	1000	10,000
509-14-8	Tetranitromethane [Methane, tetranitro-]		(b)		10,000
7719-09-7	Thionyl chloride	T		250	
7550-45-0	Titanium tetrachloride [Titanium chloride ($TiCl_4$) (T-4)-]		(b)		2500
584-84-9	Toluene 2,4-diisocyanate [Benzene, 2,4-diiso-cyanato-1-methyl-]		(a) [5]		10,000
91-08-7	Toluene 2,6-diisocyanate [benzene, 1,3-diiso-cyanato-2-methyl-]		(a) [5]		10,000
26471-62-5	Toluene diisocyanate (unspecified isomer) [Benzene, 1,3-diisocyanatomethyl-]		(a) [5]		10,000
1558-25-4	Trichloro(chloromethyl) silane	T		100	

CHEMICAL		LISTING BASIS (See Notes)		THRESHOLD QUANTITY, pounds	
CAS #	NAME	OSHA	EPA	OSHA TQ [1]	EPA TQ [2]
21737-85-5	Trichloro(dichloro-phenyl)silane	T		2500	
10025-78-2	Trichlorosilane	T	See flam-mables	5000	
79-38-9	Trifluorochloroethylene	T	See flam-mables	10,000	
75-77-4	Trimethylchlorosilane [Silane, chlorotrimethyl-]	[6]	(b)		10,000
2487-90-3	Trimethyoxysilane	T		1500	
108-05-4	Vinyl acetate monomer [Acetic acid ethenyl ester]	[6]	(b)		15,000

CHEMICAL		LISTING BASIS (See Notes)		THRESHOLD QUANTITY, pounds	
CAS #	NAME	N/A	EPA	N/A	EPA TQ [8]
EPA FLAMMABLES					
75-07-0	Acetaldehyde		(g)		10,000
74-86-2	Acetylene [Ethyne]		(f)		10,000
598-73-2	Bromotrifluorethylene [Ethene, bromotrifluoro-]		(f)		10,000
106-99-0	Butadiene, 1,3-		(f)		10,000
106-97-8	Butane		(f)		10,000
25167-67-3	Butene		(f)		10,000
106-98-9	Butene, 1-		(f)		10,000
107-01-7	Butene, 2-		(f)		10,000
590-18-1	Butene-cis, 2-		(f)		10,000
624-64-6	Butene-trans, 2- [2-Butene, (E)]		(f)		10,000

CHEMICAL		LISTING BASIS (See Notes)		THRESHOLD QUANTITY, pounds	
CAS #	NAME	N/A	EPA	N/A	EPA TQ [8]
463-58-1	Carbon oxysulfide [Carbon oxide sulfide (COS)]		(f)		10,000
7791-21-1	Chlorine monoxide [Chlorine oxide]		(f)		10,000
590-21-6	Chloropropylene, 1- [1-propene, 1-chloro-]		(g)		10,000
557-98-2	Chloropropylene, 2- [1-propene, 2-chloro-]		(g)		10,000
460-19-5	Cyanogen [Ethanedinitrile]		(f)		10,000
75-19-4	Cyclopropane		(f)		10,000
4109-96-0	Dichlorosilane [Silane, dichloro-]		(f)		10,000
75-37-6	Difluoroethane [Ethane, 1,1-difluoro-]		(f)		10,000
124-40-3	Dimethylamine [Methanamine, n-methyl-]		(f)		10,000
463-82-1	Dimethylpropane, 2,2- [Propane, 2,2-dimethyl-]		(f)		10,000
74-84-0	Ethane		(f)		10,000
107-00-6	Ethyl acetylene [1-Butyne]		(f)		10,000
75-00-3	Ethyl chloride [Ethane, chloro-]		(f)		10,000
60-29-7	Ethyl ether [ethane, 1,1'-oxybis-]		(g)		10,000
75-08-1	Ethyl mercaptan [Ethanethiol]		(g)		10,000
109-95-5	Ethyl nitrite [Nitrous acid, ethylester]		(f)		10,000
75-04-7	Ethylamine [Ethanamine]		(f)		10,000
74-85-1	Ethylene [Ethene]		(f)		10,000
1333-74-0	Hydrogen		(f)		10,000
75-28-5	Isobutane [Propane, 2-methyl]		(f)		10,000
78-78-4	Isopentane [Butane, 2-methyl-]		(g)		10,000

CHEMICAL		LISTING BASIS (See Notes)		THRESHOLD QUANTITY, pounds	
CAS #	NAME	N/A	EPA	N/A	EPA TQ [8]
78-79-5	Isoprene [1,3-Butadiene, 2-methyl-]		(g)		10,000
75-29-6	Isopropyl chloride [Propane,2-chloro-]		(g)		10,000
75-31-0	Isopropylamine [2-Propanamine]		(g)		10,000
74-82-8	Methane		(f)		10,000
563-45-1	Methyl-1-butene, 3-		(f)		10,000
563-46-2	Methyl-1-butene, 2-		(g)		10,000
115-10-6	Methyl ether [methane, oxybis-]		(f)		10,000
107-31-3	Methyl formate [Formic acid, methyl ester]		(g)		10,000
74-89-5	Methylamine [Methanamine]		(f)		10,000
115-11-7	Methylpropene, 2- [1-Propene, 2-methyl-]		(f)		10,000
504-60-9	Pentadiene 1,3-		(f)		10,000
109-66-0	Pentane		(g)		10,000
646-04-8	Pentene (E)-, 2-		(g)		10,000
627-20-3	Pentene, (Z)-, 2-		(g)		10,000
109-67-1	Pentene, 1-		(g)		10,000
463-49-0	Propadiene [1,2-Propadiene]		(f)		10,000
74-98-6	Propane		(f)		10,000
115-07-1	Propylene [1-Propene]		(f)		10,000
74-99-7	Propyne [1-Propyne]		(f)		10,000
7803-62-5	Silane		(f)		10,000
116-14-3	Tetrafluoroethylene [Ethene, tetrafluoro-]		(f)		10,000
75-76-3	Tetramethylsilane [Silane, tetramethyl-]		(g)		10,000

CHEMICAL		LISTING BASIS (See Notes)		THRESHOLD QUANTITY, pounds	
CAS #	NAME	N/A	EPA	N/A	EPA TQ [8]
10025-78-2	Trichlorosilane [Silane, trichloro-]		(g)		10,000
79-38-9	Trifluorochloroethylene [Ethene, chlorotrifluoro-]		(f)		10,000
75-50-3	Trimethylamine [Methanamine, N,N-dimethyl-]		(f)		10,000
689-97-4	Vinyl acetylene [1-Buten-3-yne]		(f)		10,000
75-01-4	Vinyl chloride [Ethene, chloro-]		(a,f)		10,000
109-92-2	Vinyl ethyl ether [Ethene, ethoxy-]		(g)		10,000
75-02-5	Vinyl fluoride [Ethene, fluoro-]		(f)		10,000
107-25-5	Vinyl methyl ether [Ethene, methoxy-]		(f)		10,000
75-35-4	Vinylidene chloride [Ethene, 1,1-dichloro-]		(g)		10,000
75-38-7	Vinylidene fluoride [Ethene, 1,1-difluoro-]		(f)		10,000

Notes:

EPA regulated toxic substances while OSHA regulated both toxic and reactive substances (those having an NFPA 49 reactivity ratings of 3 or 4). While OSHA did not delineate which substances were regulated for which property (toxicity or reactivity), reasonable inference can be made from parallels between the OSHA list and the list of substances regulated under Delaware's Extremely Hazardous Substances Risk Management regulations. This has been indicated above with a "T" for toxic and an "R" for reactive in the OSHA Listing Basis column.

The specific bases for listing of EPA substances are indicated via the following footnotes:

(a) Mandated for listing by Congress.

(b) On EHS list, vapor pressure 10 mm Hg or greater.

(c) Toxic gas.

Notes (*continued*)

(d) Toxicity of hydrogen chloride, potential to release hydrogen chloride, and history of accidents.

(e) Toxicity of sulfur trioxide, potential to release sulfur trioxide, and history of accidents.

(f) Flammable gas.

(g) Volatile flammable liquid.

Other footnotes:

(1) OSHA generally covers only commercial strengths of chemicals; dilutions of these chemicals (except for those specifically listed as solutions) are not covered. Where the listed HHC is a solution (at or above a specified minimum concentration), the total weight of the solution is the quantity that is compared to the TQ.

(2) EPA TQ's for toxics are, with the exception of oleum, intended to be applied to the mass of the regulated chemical at whatever concentration it exists in the process; for example, if the TQ is 10,000 lbs, it pertains to 10,000 lbs of the molecules of the regulated chemical. However, see the exclusion for solutions at concentrations of < 1 wt% or with partial pressure < 10 mm Hg, per § 68.115(b)(1).

The single exception to the above "rule" is oleum; since there is no "oleum molecule" the TQ of 10,000 lbs applies to the total amount *of solution*, regardless of concentration.

(3) EPA has established separate TQ's for anhydrous ammonia and for ammonia in solutions; and for anhydrous hydrogen chloride and hydrochloric acid solutions. In each case, the TQ pertains to the pounds of regulated substance, not the pounds of solution. Note that the solutions are not regulated unless the concentration is equal to or greater than 20 weight percent for ammonia or 37 weight percent for hydrochloric acid.

(4) Methyl chloride is the only substance for which the EPA threshold quantity is smaller than the OSHA threshold quantity.

(5) The mixture exemption in §68.115(b)(1) does not apply to this substance

(6) These substances, while not specifically listed by OSHA, would be covered by OSHA as flammables.

(7) EPA increased the solution concentration cut-off from 30 to 37% in a Final Rule issued 8/25/97.

(8) See §68.115(b)(2), as revised by the 1/6/98 Final Rule, for EPA's handling of TQ's for flammable mixtures.

Appendix E

Example RMPlan—Propane Industry

Chemical Emergency Preparedness and Prevention Office (CEPPO) Sample Risk Management Plan under the Clean Air Act Amendments, Section 112(r)

A sample RMP (including both the Executive Summary and Data Elements) for a propane facility follows. This sample is NOT the Model RMP for the propane industry, and does NOT represent the electronic or paper format for RMP's that EPA will accept. The official reporting format is currently being developed.

EXECUTIVE SUMMARY

1. **Accidental release prevention and emergency response policies:**

In this distribution facility, we handle propane which is considered hazardous by EPA. The same properties that makes propane valuable as a fuel also makes it necessary to observe certain safety precautions in handling propane to prevent unnecessary human exposure, to reduce the threat to our own personal health as well as our co-workers, and to reduce the threat to nearby members of the community. It is our policy to adhere to all applicable Federal and state rules and regulations. Safety depends upon the manner in which we handle propane combined with the safety devices inherent in the design of this facility combined with the safe handling procedures that we use and the training of our personnel.

Our emergency response program is based upon the NPGA's LP-Gas Safety Handbook, "Guidelines for Developing Plant Emergency Procedures" and "How to Control LP-Gas Leaks and Fires." The emergency response plan includes procedures for notification of the local fire authority and notification of any potentially affected neighbors.

2. **The stationary source and regulated substances handled.**

The primary purpose of this facility is to repackage and distribute propane to both retail and wholesale customers. Propane is used by our retail customers as a fuel. Propane is received by rail car and by truck (transports) and stored in three storage tanks. Propane is distributed to retail customers by delivery trucks (bobtails) and to wholesale customers by bobtails and transports. We also fill Department of Transportation (DOT) containers for use by retail customers. This facility has equipment for unloading rail cars and transports and equipment to load bobtails, transports and DOT containers. Access to the site is restricted to authorized facility employees, authorized management personnel and authorized contractors. The regulated substance handled at this distribution facility is propane. The maximum amount of propane that can be stored at this plant is 400,000 pounds.

3. **The worst-case release scenario(s) and the alternative release scenario(s), including administrative controls and mitigation measures to limit the distances for each reported scenario.**

Worst-Case Scenario. Failure of my largest storage tank when filled to the greatest amount allowed would release 222,000 pounds of propane. Company policy limits the maximum filling capacity of this tank to 88% at 60 F. It is assumed that the entire contents are released as vapor which finds an ignition source, 10% of the released quantity is assumed to participate in the resulting explosion. The distance to the endpoint of 1 psi for the worst-case scenario is 0.47 miles.

Alternative Scenario. A pull-away causing failure of a 25 foot length of 4 inch hose. The excess flow valves function to stop the flow. The contents of the hose is released. The resulting unconfined vapor travels to the lower flammability limit. The distance to the endpoint for the lower flammability limit for the alternative scenario is less than 317 feet. This release has the possibility of extending beyond the facility boundary.

4. **The general accidental release prevention program and the specific prevention steps.**

This distribution facility complies with EPA's Accidental Release Prevention Rule and with all applicable state codes and regulations. This facility was designed and constructed in accordance with NFPA-58. All of our drivers have been thoroughly trained using the NPGA's Certified Employee Training Program (CEPT).

5. Five-year accident history.

We had an accidental release of propane that ignited on 1/28/95. No one off-site was injured and no off-site damage occurred but the adjacent highway was closed.

6. The emergency response program.

This facility's emergency response program is based upon the NPGA's LP-Gas Handbook, "Guidelines for Developing Plant Emergency Response Procedures" and "How to Control LP-Gas Leaks and Fires." We have discussed this program with the New Castle County Local Emergency Planning Committee and the Newark Fire Department. A representative of the Newark Fire Department visited this plant on June 25, 1994.

7. Planned changes to improve safety.

This facility was constructed in 1968 and is in compliance with the NFPA-58 Standard, 1967 Edition. In 2000 we plan to do extensive maintenance and upgrade the facility to NFPA-58, 1998 Edition.

RISK MANAGEMENT PLAN DATA ELEMENTS

1. REGISTRATION

1.1 Source identification:

a. **Name:** Super Safe Propane Distributors

b. **Street:** Propane Court

c. **City:** Their Fair City

d. **County:** County name

e. **State:** ST

f. **Zip:** 55555

g. **Latitude:** 7705'00"

h. **Longitude:** 3852'30"

1.2 Source Dun and Bradstreet number: 12-3456-7899

1.3

a. **Name of corporate parent company (if applicable):** N/A

b. **Dun and Bradstreet number of corporate parent company (if applicable):** N/A

1.4 Owner/operator:

a. **Name:** Firstname Lastname

b. **Phone:** (555) 555-5555

c. **Mailing address:** P.O. Box 5555, Their Fair City, ST 55555

1.5 Name and title of person responsible for part 68 implementation:
Firstname M. Lastname

1.6 Emergency contact:

a. **Name:** Firstname M. Lastname

b. **Title:** Vice President

c. **Phone:** (555) 555-5555

d. **24-hour phone:** (555) 555-1212

1.7 For each covered process:

a. **1. Chemical name: 2. CAS number: 3. Quantity: 4. SIC code: 5. Program level:**

Propane: 74-98-6: 400,000: 5172: Level 2

b. **1. Chemical name: 2. CAS number: 3. Quantity 4. SIC code: 5. Program level:**

c. **1. Chemical name: 2. CAS number: 3. Quantity 4. SIC code: 5. Program level:**

1.8 EPA Identifier: To be determined

1.9 Number of full-time employees: 5

1.10 Covered by:

a. **OSHA PSM:** 1.___Yes 2._X_No

b. **EPCRA section 302:** 1._X_Yes 2.___No

c. **CAA Title V operating permit:** 1.___Yes 2._X_No

1.11 Last safety inspection: Date: By:

a. 3/20/95

b. __OSHA

c. __State OSHA

d. __EPA

e. _X_State EPA

f. __Fire department

g. __Other (specify)

h. __Not applicable

4. WORST CASE (FLAMMABLES)

(Note: This section will be replaced once EPA has finalized the submittal format.)

4.1 Chemical: Propane

4.2 Results based on (check one):

a. _X_Reference Table

b. ___Modeling

c. ___Model used ____________________

4.3 Scenario (check one):

a. _X_Vapor cloud explosion

b. ___Fireball

c. ___BLEVE

d. ___Pool Fire

e. ___Jet Fire

4.4 Quantity released: 222,000 lbs

4.5 Endpoint used: 1 psi

4.6 Distance to endpoint: 0.47 miles

4.7 Population within distance: 25

4.8 Public receptors (check all that apply):

a. ___Schools

b. _X_Residences

c. ___Hospitals

d. ___Prisons

e. ___Public recreational areas or arenas

f. ___Major commercial or industrial areas

4.9 Environmental receptors within distance (check all that apply):

a. ___National or state parks, forests, or monuments

b. ___Officially designated wildlife sanctuaries, preserves, or refuges

c. ___Federal wilderness areas

4.10 Passive mitigation considered (check all that apply):

a. ___Dikes

b. ___Fire Walls

c. ___Blast Walls

d. ___Enclosures

e. ___Other (specify)

5. ALTERNATIVE SCENARIOS (FLAMMABLES)

(Note: This section will be replaced once EPA has finalized the submittal format.)

5.1 Chemical: Propane

5.2 Results based on (check one):

a. _X_Reference Table

b. ___Modeling

c. ___Model used ____________________

5.3 Scenario (check one):

a. _X_Vapor cloud explosion

b. ___Fireball

c. ___BLEVE

d. ___Pool Fire

e. ___Jet Fire

5.4 Quantity released: 69 lbs

5.5 Endpoint used: LFL

5.6 Distance to endpoint: .06 miles

5.7 Population within distance: 0

5.8 Public receptors (check all that apply):

a. ____Schools
b. ____Residences
c. ____Hospitals
d. ____Prisons
e. ____Public recreational areas or arenas
f. ____Major commercial or industrial areas

5.9 Environmental receptors within distance (check all that apply):

a. ____National or state parks, forests, or monuments
b. ____Officially designated wildlife sanctuaries, preserves, or refuges
c. ____Federal wilderness areas

5.10 Passive mitigation considered (check all that apply):

a. ____Dikes
b. ____Fire Walls
c. ____Blast Walls

5.11 Active mitigation considered (check all that apply):

a. ____Sprinkler Systems
b. ____Deluge Systems
c. ____Water Curtain
d. _X_Excess Flow Valve

6. FIVE-YEAR ACCIDENT HISTORY (complete for each release)

(Note: This section will be replaced once EPA has finalized the submittal format.)

6.1 Date: 1/28/95

6.2 Time: 8:05 AM

6.3 Release duration: 10 Minutes

6.4 Chemical(s): Propane

6.5 Quantity released (lbs): 50

6.6 Release event:

6.7 Release/source:

a. _X_ Gas release

b. ___ Liquid spill/evaporation

c. _X_ Fire

d. ___ Explosion

e. ___ Valve

f. _X_ Pump

a. ___ Storage vessel

b. ___ Piping

c. ___ Process vessel

d. _X_ Transfer hose

6.8 Weather conditions at time of event (if known):

a. Wind speed/direction________

b. Temperature________

c. Stability class________

d. Precipitation present___No___

e. Unknown___X___

6.9 On-site impacts:

a. Deaths ______ (number)

b. Injuries ______ (number)

c. Property damage___$325___

6.10 Known offsite impacts:

a. Deaths __0__ (number)

b. Hospitalizations __0__ (number)

c. Other medical treatment __0__ (number)

d. Evacuated __0__ (number)

e. Sheltered __0__ (number)

f. Property damage ($) __0__

g. Environmental damage __None__ (specify type)

6.11 Initiating event:

6.12 Contributing factors (check all that apply):

a. ____Equipment failure

b. _X_Human error

c. ____Improper procedures

d. ____Overpressurization

e. ____Upset condition

f. ____By-pass condition

g. ____Maintenance activity/Inactivity

h. ____Process design

i. ____Unsuitable equipment

j. ____Unusual weather condition

k. ____Management error

6.13 Offsite responders notified

a. _X_Yes

b. ____No

6.14 Changes introduced as a result of the accident

a. ____Improved/upgraded equipment

b. _X_Revised maintenance

c. ____Revised training

d. ____Revised operating procedures

e. ____New process controls

f. ____New mitigation systems

g. _X_Revised emergency response plan

h. ____Changed process

i. ____Reduced inventory

j. ____Other

k. ____ None

8. PREVENTION PROGRAM 2 (For Each Program 2 Process)

(Note: This section will be replaced once EPA has finalized the submittal format.)

8.1 SIC code for process: 5172

8.2 Chemicals:

a. Propane

8.3 Safety information:

a. The date of the most recent review or revision of the safety information: 5/23/96

b. A list of Federal or state regulations or industry-specific design codes and standards used to demonstrate compliance with safety information requirement:

1. _X_ NFPA 58 (or state law based on NFPA 58)
2. _X_ OSHA 1910.111
3. ____ ASTM
4. _X_ ANSI standards
5. _X_ ASME standards
6. ____ other (specify)
7. ____ None

8.4 Hazard Review:

a. The date of completion of the most recent hazard review or update: 5/24/96

b. The expected date of completion of any changes resulting from the hazard review: 5/24/96

c. Major hazards identified (check all that apply):

1. ____ Toxic release
2. _X_ Fire
3. _X_ Explosion
4. ____ Runaway reaction
5. ____ Polymerization
6. ____ Overpressurization

7. ____Corrosion
8. _X_Overfilling
9. ____Contamination
10. _X_Equipment failure
11. ____Loss of cooling, heating, electricity, instrumentation air
12. ____Earthquake
13. ____Floods (flood plain)
14. ____Tornado
15. ____Hurricanes
16. ____Other

d. Process controls in use (check all that apply):

1. ____Vents
2. _X_Relief valves
3. ____Check valves
4. ____Scrubbers
5. ____Flares
6. _X_Manual shutoffs
7. _X_Automatic shutoffs
8. ____Interlocks
9. ____Alarms and procedures
10. ____Keyed bypass
11. ____Emergency air supply
12. ____Emergency power
13. ____Backup pump
14. _X_Grounding equipment
15. ____Inhibitor addition
16. ____Rupture disks
17. _X_Excess flow device
18. ____Quench system
19. ____Purge system
20. ____Other

e. Mitigation systems in use (check all that apply)

1. ____Sprinkler system
2. ____Dikes
3. ____Fire walls
4. ____Blast walls
5. ____Deluge systems
6. ____Water curtain
7. ____Enclosure
8. ____Neutralization
9. ____Other

f. Monitoring/detection systems in use

1. ____Process area detectors
2. ____Perimeter monitors
3. ____Other

g. Changes since last hazard review update (check all that apply)

1. ____Reduction in chemical inventory
2. ____Increase in chemical inventory
3. ____Change in process parameters
4. ____Installation of process controls
5. ____Installation of process detections systems
6. ____Installation of perimeter monitoring systems
7. ____Installation of mitigating systems
8. ____Other
9. _X_None required recommended

8.5 The date of the most recent review or revision of operating procedures: 6/4/96

8.6 Training:

a. The date of the most recent review or revision of training programs: 6/5/96

b. The type of training provided:

1. ____Classroom
2. _X_Classroom plus on the job

3. ____On the job
4. ____Other

c. The type of competency testing used:

1. _X_Written tests
2. ____Oral tests
3. _X_Demonstration
4. _X_Observation
5. ____Other

8.7 Maintenance:

a. The date of the most recent review or revision of maintenance procedures: 6/15/96

b. The date of the most recent equipment inspection or test: 6/16/96

c. The equipment inspected or tested: Pump

8.8 Compliance audits:

a. The date of the most recent compliance audit: 6/30/96

b. The expected date of completion of any changes resulting from the compliance audit: 6/30/96

8.9 Incident investigation:

a. The date of the most recent incident investigation: 1/28/95

b. The expected date of completion of any changes resulting from the investigation: 3/30/95

8.10 The date of the most recent change that triggered a review or revision of safety information, the hazard review, operating or maintenance procedures, or training: 6/16/96

9. EMERGENCY RESPONSE

(Note: This section will be replaced once EPA has finalized the submittal format.)

9.1 Do you have a written emergency response plan?:

a. _X_Yes

b. ____No

9.2 Does the plan include specific actions to be taken in response to an accidental release of a regulated substance?:

a. _X_ Yes

b. ___No

9.3 Does the plan include procedures for informing the public and local agencies responsible for responding to accidental releases?:

a. _X_ Yes

b. ___No

9.4 Does the plan include information on emergency health care?:

a. _X_ Yes

b. ___No

9.5 The date of the most recent review or update of the emergency response plan: 2/28/96

9.6 The date of most recent emergency response training for employees: 3/10/96

9.7 The name and telephone of the local agency with which the plan is coordinated:

a. **Name:** Newark Fire Department

b. Telephone number: (302) 888-1212 or 311

9.8 Subject to (check all that apply):

a. ___OSHA 1910.38 (Emergency Action Plan)

b. ___OSHA 1910.120 (HAZWOPER)

c. ___Clean Water Act / SPCC

d. ___RCRA

e. ___OPA-90

f. ___State EPCRA Rules / Law

g. ___Other (specify)

References

1. *A Compliance Guideline for EPA's Risk Management Program Rule,* The Chemical Manufacturers Association and the American Petroleum Institute, 1997.
2. *A Model Risk Management Plan for E&P Facilities,* American Petroleum Institute, 1997.
3. *A Model Risk Management Plan for Refineries,* American Petroleum Institute, 1997.
4. Accidental Release Prevention Requirements: Risk Management Programs Under Clean Air Act Section 112(r)(7), *Federal Register,* Vol. 61, No. 120, p. 31667, June 20, 1996.
5. Extremely Hazardous Substances Risk Management Act, Delaware Code, Title 7, Chapter 77, 1989.
6. Hazardous Materials: Accidental Release, California Senate Bill No. 1889, September 1996.
7. List of Regulated Substances and Thresholds for Accidental Release Prevention; Requirements for Petitions Under Section 112(r)(7) of the Clean Air Act as Amended, *Federal Register,* Vol. 59, No. 20, p. 4478, January 31, 1994.
8. List of Regulated Substances and Thresholds for Accidental Release Prevention: Final Rule, *Federal Register,* Vol. 63, No. 3, p. 639, January 6, 1998.
9. *Management of Process Hazards,* API Recommended Practice 750, First Edition, American Petroleum Institute, Washington, DC, January 1990.
10. Process Safety Management of Highly Hazardous Chemicals, *Federal Register,* Vol. 57, No. 36, p. 6356, February 24, 1992.
11. Resource Conservation and Recovery Act Contingency Planning Requirements, 40 CFR Part 264, Subpart D, 40 CFR Part 265, Subpart D, 40 CFR 279.52, Environmental Protection Agency.
12. *Responsible Care,* Chemical Manufacturers Association, 1988.
13. RMP Data Elements Checklist, Docket A-91-73 Category VIII-A, Environmental Protection Agency, Washington, DC, May 24, 1996.
14. RMP Data Elements Instructions, Docket A-91-73 Category VII-A, Environmental Protection Agency, Washington, DC, May 24, 1996.

15. RMP Offsite Consequence Analysis Guidance, Docket A-91-73 Category VIII-A, Environmental Protection Agency, Washington, DC, May 24, 1996.
16. *Strategies for Today's Environmental Partnership*, American Petroleum Institute, Washington, DC, 1990.
17. The National Response Team's Integrated Contingency Plan Guidance, *Federal Register*, Vol. 61, No. 109, June 5, 1996.
18. The Role of Local Emergency Planning Committees (LEPCs) and Other Local Agencies in the Risk Management Program (RMP) of Clean Air Act (CAA) Section 112(r)—Subgroup #7 Report (CEPPO), Office of Solid Waste and Emergency Response (OSWER), U.S. Environmental Protection Agency (EPA)

Glossary

Accidental release. An unanticipated emission of a regulated substance or other extremely hazardous substance into the ambient air from a stationary source.

Acute exposure. Refers to a single exposure that occurs over a relatively short period of time (for example, the type of exposure from a vapor cloud resulting from an accidental release).

Administrative controls. Written procedural mechanisms used for hazard control.

Alternative release scenarios. The scenarios other than worst case provided in the hazard assessment. For these scenarios, facilities may consider the effects of both passive and active mitigation systems.

Article. A manufactured item, as defined under 29 CFR 1910.1200(b), that is formed to a specific shape or design during manufacture, that has end use functions dependent in whole or in part upon the shape or design during end use, and that does not release or otherwise result in exposure to a regulated substance under normal conditions of processing and use.

Atmospheric stability. A classification of the amount of turbulence (horizontal and vertical movement of the surrounding air) that exists in the atmosphere at any given time. Levels of atmospheric stability are identified with a letter (A - F). Unstable conditions (A - C) generally occur during mid-day with clear skies and light winds; these conditions cause considerable horizontal and vertical turbulence and result in rapid dispersion of a vapor cloud as it moves downwind. Neutral conditions (D) can occur during the day or night with cloudy skies and moderate-to-strong winds; these conditions cause less turbulence in the horizontal and vertical directions than unstable conditions and result in less rapid dispersion of the vapor cloud as it moves downwind. Stable conditions (E - F) generally occur at night or early morning with clear skies and light winds; there is very little horizontal or vertical turbulence, which results in very slow dispersion of the vapor cloud as it moves downwind.

Boiling liquid expanding vapor explosion (BLEVE). The explosive vaporization of a superheated liquid when it is instantaneously released from a vessel. The resulting release of energy generates an overpressure, and a fireball often occurs if the material is combustible and the container/vessel failure is caused by an external fire. The three main consequences of a BLEVE are (1) the overpressure that could be generated, (2) vessel fragments that may be propelled away from the explosion, and (3) when applicable, thermal radiation from the fireball.

Catastrophic release. An uncontrolled emission, fire, or explosion, involving one or more regulated substances that presents imminent and substantial endangerment to public health and the environment.

Classified Information. Defined in the Classified Information Procedures Act, 18 U.S.C. App. 3, Section 1(a) as "any information or material that has been determined by the United States Government pursuant to an executive order, statute, or regulation, to require protection against unauthorized disclosure for reasons of national security."

Concentration in air, parts per million (ppm), % by volume (vol%). The relative amount by volume of a material contained within a vapor cloud in the air, often expressed in parts per million (ppm) or % by volume (vol%). A concentration of 1,000,000 ppm (or 100 vol%) means that the vapor cloud volume consists only of the material with no air. A concentration of 500,000 ppm (or 50 vol%) means that the vapor cloud volume is one-half material and one-half air.

Condensate. Hydrocarbon liquid separated from natural gas that condenses due to changes in temperature, pressure, or both, and remains liquid at standard conditions.

Consequence analysis. The predicted effects of accidental releases using mathematical models, historical experience of accident effects, or experimental results. It typically Includes estimating a source term, predicting the transport of energy or the release of material through the environment, and estimating the effects of the release.

Covered process. A process that has a regulated substance present in more than a threshold quantity as determined under §68.115 of 40 CFR 68.

Crude oil. Any naturally occurring, unrefined petroleum liquid.

Designated agency. Any state, local, or federal agency designated by the state as the agency responsible for the review of an RMP for completeness.

Dispersion model. Any method used to predict the characteristics of a vapor cloud as it moves downwind. It is typically based upon release information and meteorological data to determine vapor cloud concentration and dimensions. The method may be based on experimental data, theoretical data, or a combination of the two.

Endpoint. A toxic substance's Emergency Response Planning Guideline level 2 (ERPG 2) developed by the American Industrial Hygiene Association (AIHA). If a substance has no ERPG 2, then the endpoint is the level of concern (LOC) from the *Technical Guidance for Hazards Analysis*, updated where necessary to reflect new toxicity data. For vapor cloud explosions, 1-psi overpressure and for fires (that is, jet fires, pool fires, fireballs), 5 kw/m^2 for 40 seconds, are to be used. For vapor cloud fires and jet fires, the lower flammability limit provided by the NFPA or other sources shall be used.

Emergency response planning guideline (ERPG). The concentration of a hazardous material in air above which some members of the public may begin to experience adverse effects. The AIHA approves and publishes three levels (ERPG 1, ERPG 2, and ERPG 3, defined below), each related to the severity of effect.

Environmental receptor. Natural areas such as national or state parks, forests, or monuments; officially designated wildlife sanctuaries, preserves, refuges, or areas; and federal wilderness areas, that could be exposed at any time to toxic concentrations, radiant heat, or overpressure greater than or equal to the endpoints provided in □68.22(a) of 40 CFR 68, as a result of an accidental release and that can be identified on local U.S. Geological Survey maps.

ERPG 1. The maximum airborne concentration below which it is believed nearly all individuals could be exposed for up to 1 hour without experiencing other than mild transient adverse health effects or perceiving a clearly defined objectionable odor.

ERPG 2. The maximum airborne concentration below which it is believed nearly all individuals could be exposed for up to 1 hour without experiencing or developing irreversible or other serious health effects or symptoms that could impair their abilities to take protective action.

ERPG 3. The maximum airborne concentration below which it is believed nearly all individuals could be exposed for up to 1 hour without experiencing or developing life-threatening health effects.

Explosion. A release of energy that causes a transient change in the density, pressure, and velocity of the air surrounding the source of energy. This release of energy may generate a damaging pressure wave. If the source of energy originates from rapid depressurization of a vessel (as in a pressure vessel rupture or BLEVE), this is referred to as a *physical explosion*: If the source of energy originates from combustion of flammable material (as in a vapor cloud explosion), it is called a *chemical explosion*.

Exposure time. The total time interval over which an individual is actually exposed to a hazardous condition (material in a vapor cloud, fire, etc.).

Field gas. Gas extracted from a production well before the gas enters a natural gas processing plant.

Fireball. A fireball results following immediate ignition of an instantaneous release of a flammable vapor or superheated liquid or liquid and vapor mixture. The burning cloud tends to rise, expand, and assume a spherical shape. A fireball usually exists for only 10 to 20 seconds; however, it may present thermal radiation effects and severely burn individuals hundreds of feet from the source of the fireball. A fireball often accompanies a BLEVE if the released liquid is flammable and the release results from vessel failure caused by an external fire.

Hazard assessment. As used in EPA's RMP rule, an analysis to estimate the potential consequences of accidental releases of hazardous materials on the public and on the environment.

Implementing agency. The state or local agency that obtains delegation for an accidental release prevention program under subpart E of part 63 under section 112(l) of the CAA. The implementing agency may, but is not required to, be the state or local air permitting agency. If a state or local agency does not take delegation, EPA will be the implementing agency for the state.

Injury. Any effect on a human that results either from direct exposure to toxic concentrations; radiant heat; or overpressures from accidental releases or from the direct consequences of a vapor cloud explosion (such as flying glass, debris, and other projectiles) from an accidental release and that requires medical treatment or hospitalization.

Jet fire. These result from the ignition of a flammable vapor or liquid/vapor mixture that is being continuously discharged from an orifice, leak, or rupture. The resulting flame has a torch-like appearance and may pose thermal radiation hazards to nearby individuals.

Local emergency planning committee (LEPC). A local interdisciplinary group appointed by the State Emergency Response Commission (SERC) to develop a comprehensive emergency plan for responding to accidental releases of hazardous materials that could affect the public. Individual plants/facilities have the primary responsibility of responding to onsite emergencies, while the LEPC is responsible for developing plans for safeguarding the public if hazardous materials migrate off site. The membership of the LEPC must include local citizens, emergency responders, members of law enforcement, local media, as well as industry representatives.

Major change. Introduction of a new process, process equipment, or regulated substance, an alteration of process chemistry that results in any change to safe operating limits, or other alteration that introduces a new hazard.

Medical treatment. Treatment, other than first aid, administered by a physician or registered professional personnel under standing orders from a physician.

Mitigation system, active, passive. Specific activities, technologies, or equipment designed or deployed to capture or control substances upon loss of containment to minimize exposure of the public or the environment. Passive mitigation is equipment, devices, or technologies that function without human, mechanical, or other energy input. An example of a passive mitigation system is a dike surrounding a storage tank that limits the spread and vaporization of a spilled hazardous material. Active mitigation is equipment, devices, or technologies that need human, mechanical, or other energy input to function. An example of an active mitigation system is an automatic shutoff valve that limits the duration of a hazardous material release.

Natural gas processing plant (gas plant). Any processing site engaged in the extraction of natural gas liquids from field gas, fractionation of mixed natural gas liquids to natural gas products, or both. A separator, dehydration unit, heater treater, sweetening unit, compressor, or similar equipment shall not be considered a "processing site" unless such equipment is physically located within a natural gas processing plant (gas plant) site.

Off site. Areas beyond the property boundary of the stationary source or areas within the property boundary to which the public has routine and unrestricted access during or outside business hours.

Overpressure. The sudden increase in the local atmospheric pressure that may result from an explosion. The standard pressure in the atmosphere is approximately 14.7 pounds per square inch at sea level. An explosion that causes a 3 lb per square inch overpressure means that the local atmospheric pressure suddenly increased from 14.7 to 17.7 lb per square inch. Significant overpressure may cause severe injury to exposed individuals and damage to property.

Pool depth. The thickness of a liquid pool that is spilled onto a given surface (concrete, gravel, soil, water, etc.). The minimum pool depth that a liquid spill may attain as it spreads out depends on such factors as the roughness and contour of the surface, the liquid viscosity, and the liquid pour point temperature.

Pool fire. Results from the ignition of flammable vapors that evaporate from a flammable liquid spill. The flames associated with the pool fire may produce thermal radiation effects to individuals located near the fire.

Population. The public.

Pressure wave. A moving disturbance that emanates from an explosion and causes a localized increase in atmospheric pressure (overpressure) as it traverses the atmosphere.

Process. Any activity involving a regulated substance, including any use, storage, manufacturing, handling, or the onsite movement of such chemicals or combination of these activities. For purposes of this definition, any group of vessels that are interconnected and separate vessels that are located such that a highly hazardous chemical could be involved in a potential release shall be considered a single process.

Public. Any person except employees or contractors at the stationary source.

Public receptor. Offsite residences; institutions (for example, schools, hospitals), industrial, commercial, and office buildings; parks, or recreational areas inhabited or occupied by the public at any time without restriction by the stationary source where members of the public could be exposed to toxic concentrations, radiant heat, or overpressure, as a result of an accidental release.

Regulated substance. Any substance listed pursuant to section 112 (r) (3) of the CAA as amended in §68.130 of 40 CFR 68.

Release duration. The total time interval over which a hazardous material is being released to the surrounding air.

Release rate. Refers to the quantity (in pounds, kilograms, gallons, or other measurement) of a hazardous material that is released per unit time (per second, per minute, per hour) from a tank, pipe, or other piece of equipment.

Shelter-in-place. A method of protecting oneself from exposure to a toxic vapor cloud by remaining inside an enclosure (building or house) until the concentration within the vapor cloud (outside of the enclosure) has decreased to a safe level.

Source term. Defines the quantity or release rate, the duration of the release, and the form (liquid, vapor, or liquid and vapor) for an accidental release of a hazardous material.

Stationary source. Any buildings, structures, equipment, installations, or substance emitting stationary activities, which belong to the same industrial group, which are located on one or more contiguous properties, which are under the control of the same person (or persons under common control), and from which an accidental release may occur. A stationary source includes transportation containers that are no longer under active shipping papers and transportation containers that are connected to equipment at the stationary source for the purposes of temporary storage, loading, or unloading. The term stationary source does not apply to transportation, including the storage incident to transportation, of any regulated substance or any other extremely hazardous substance under the provisions of this Part, provided that such transportation is regulated under 49 CFR part 192, 193, or 195, or a state natural gas or hazardous liquid program for which the state has in effect a certification to DOT under 49 U.S.C. Section 60105. Properties shall not be considered contiguous solely because of a railroad or gas pipeline right-of-way.

Surface roughness. A measure of the weighted-average height of surface objects (grass, trees, buildings, etc.) in the vicinity (upwind and downwind) of the released hazardous material. The surface roughness influences the atmospheric dispersion of a released hazardous material by increasing turbulence (horizontal and/or vertical movement) of the surrounding air. Small values of surface roughness create less turbulence and result in less rapid dilution of the cloud as it moves downwind, while larger values of surface roughness create more turbulence and result in more rapid dilution of the cloud as it moves downwind.

Threshold quantity. The quantity specified for regulated substances pursuant to section 112®(5) of the CAA as amended, listed in §68.130 and determined to be present at a stationary source as specified in §68.115 of 40 CFR 68.

Thumbprint. The area potentially affected by an accidental release of hazardous material in which the level of concern is exceeded. For example, the thumbprint for a toxic release could represent the area covered by the toxic cloud in which the average concentration of the material in the cloud exceeded the ERPG 3 value. For an explosion, the thumbprint would be the area in which the level of concern for overpressure would be exceeded. (See "Vulnerability zone.")

Typical meteorological conditions. The temperature, wind speed, cloud cover, and atmospheric stability class prevailing at the site, based on data gathered at or near the site or from a local meteorological station.

Vapor cloud explosion (VCE). Results from the ignition of a cloud of flammable vapor or vapor/mist. The burning cloud generates expanding gases so quickly that a damaging pressure wave is produced. Partial confinement and/or significant congestion, resulting in increased turbulence in the burning cloud, are usually required for high velocity flame propagation (which generates damaging overpressures). The overpressure produced by the VCE can cause severe injuries and damage at significant distances from the point of release and/or the point of ignition.

Vessel. Any reactor, tank, drum, barrel, cylinder, vat, kettle, boiler, pipe, hose, or other container.

Vulnerability zone. The vulnerability zone is the overlay of all thumbprints associated with a hypothetical accidental release of hazardous material, accounting for the variation in the wind direction at the time of the release. For a toxic release, the vulnerability zone is obtained by rotating the thumbprint to include all possible wind directions, which results in a circular area.

Wind speed. The velocity of the wind as it moves through the atmosphere, generally measured by the NWS at a height of 10 meters (33 ft) from the ground and reported based on the direction the wind is originating (for example, winds from the southeast). The wind speed is most often reported as being within some range of values (that is, 5–10 mph). The wind speed influences the atmospheric dispersion of hazardous vapor clouds. While the NWS reports wind speeds at a height of 10 meters from the ground, the wind speed does vary as a function of elevation. Wind speeds used in dispersion models should represent values that are consistent with the actual height of the release or the depth of the vapor cloud, as appropriate.

Worst-case release. Release of the largest quantity of a regulated substance from a vessel or process line failure that results in the greatest distance to an endpoint defined in §68.22(a) of 40 CFR 68.

Worst-case scenario. An accidental release involving a hazardous material that would result in the worst (most severe) off-site consequences.

Publications Available from the
CENTER FOR CHEMICAL PROCESS SAFETY
of the
AMERICAN INSTITUTE OF CHEMICAL ENGINEERS
3 Park Avenue, New York, NY 10016-5901

CCPS Guidelines Series

Guidelines for Pressure Relief and Effluent Handling Systems
Guidelines for Design Solutions for Process Equipment Failures
Guidelines for Safe Warehousing of Chemicals
Guidelines for Postrelease Mitigation in the Chemical Process Industry
Guidelines for Integrating Process Safety Management, Environment, Safety, Health, and Quality
Guidelines for Use of Vapor Cloud Dispersion Models, Second Edition
Guidelines for Evaluating Process Plant Buildings for External Explosions and Fires
Guidelines for Writing Effective Operations and Maintenance Procedures
Guidelines for Chemical Transportation Risk Analysis
Guidelines for Safe Storage and Handling of Reactive Materials
Guidelines for Technical Planning for On-Site Emergencies
Guidelines for Process Safety Documentation
Guidelines for Safe Process Operations and Maintenance
Guidelines for Process Safety Fundamentals in General Plant Operations
Guidelines for Chemical Reactivity Evaluation and Application to Process Design
Tools for Making Acute Risk Decisions with Chemical Process Safety Applications
Guidelines for Preventing Human Error in Process Safety
Guidelines for Evaluating the Characteristics of Vapor Cloud Explosions, Flash Fires, and BLEVEs
Guidelines for Implementing Process Safety Management Systems
Guidelines for Safe Automation of Chemical Processes
Guidelines for Engineering Design for Process Safety
Guidelines for Auditing Process Safety Management Systems
Guidelines for Investigating Chemical Process Incidents
Guidelines for Hazard Evaluation Procedures, Second Edition with Worked Examples
Plant Guidelines for Technical Management of Chemical Process Safety, Revised Edition
Guidelines for Technical Management of Chemical Process Safety

Guidelines for Chemical Process Quantitative Risk Analysis
Guidelines for Process Equipment Reliability Data with Data Tables
Guidelines for Safe Storage and Handling of High Toxic Hazard Materials
Guidelines for Vapor Release Mitigation

CCPS Concepts Series

Local Emergency Planning Committee Guidebook: Understanding the EPA Risk Management Program Rule
Inherently Safer Chemical Processes. A Life-Cycle Approach
Contractor and Client Relations to Assure Process Safety
Understanding Atmospheric Dispersion of Accidental Releases
Expert Systems in Process Safety
Concentration Fluctuations and Averaging Time in Vapor Clouds

Proceedings and Other Publications

Proceedings of the International Conference and Workshop on Reliability and Risk Management, 1998
Proceedings of the International Conference and Workshop on Risk Analysis in Process Safety, 1997
Proceedings of the International Conference and Workshop on Process Safety Management and Inherently Safer Processes, 1996
Proceedings of the International Conference and Workshop on Modeling and Mitigating the Consequences of Accidental Releases of Hazardous Materials, 1995
Proceedings of the International Symposium and Workshop on Safe Chemical Process Automation, 1994
Proceedings of the International Process Safety Management Conference and Workshop, 1993
Proceedings of the International Conference on Hazard Identification and Risk Analysis, Human Factors, and Human Reliability in Process Safety, 1992
Proceedings of the International Conference and Workshop on Modeling and Mitigating the Consequences of Accidental Releases of Hazardous Materials, 1991
Safety, Health and Loss Prevention in Chemical Processes: Problems for Undergraduate Engineering Curricula

Printed and bound by CPI Group (UK) Ltd, Croydon, CR0 4YY

23/04/2025

14660907-0001